# Studies in Choice and Welfare



For further volumes:
http://www.springer.com/series/6869

Bezalel Peleg • Hans Peters

# Strategic Social Choice

## Stable Representations of Constitutions

Springer

Professor Dr. Bezalel Peleg
The Hebrew University of Jerusalem
Institute of Mathematics and
Center for the Study of Rationality
Feldman Building, Givat-Ram
Jerusalem 91904
Israel
pelegba@math.huji.ac.il

Professor Dr. Hans Peters
Department of Quantitative Economics
University of Maastricht
P.O.Box 616
6200 MD Maastricht
The Netherlands
h.peters@maastrichtuniversity.nl

Studies in Choice and Welfare ISSN 1614-0311
ISBN 978-3-642-13874-4 e-ISBN 978-3-642-13875-1
DOI 10.1007/978-3-642-13875-1
Springer Heidelberg Dordrecht London New York

Library of Congress Control Number: 2010934384



*Cover design*: WMXDesign GmbH, Heidelberg

Printed on acid-free paper

Springer is part of Springer Science+Business Media (www.springer.com)

# Preface

This book was written mainly during the Spring periods of 2008 and 2009, when the first author was visiting Maastricht University. Financial support both from the Dutch Science Foundation NWO (grants 040.11.013 and 0.40.11.082) and from the research institute METEOR (Maastricht University) is gratefully acknowledged.

Jerusalem
Maastricht
April 2010

*Bezalel Peleg*
*Hans Peters*

# Contents

# Preview to this book

A *constitution* describes the *rights* and *obligations* of individuals and groups in a society. Thus, it implies to which social states or sets of social states these groups are entitled. In order to enforce such rights and obligations, laws and rules are required that set bounds to the behavior of individuals and groups: they restrict the choices that individuals can make. A few natural desiderata for such a system of laws and rules come to mind. First, these laws and rules should leave the society members enough space to be able to enforce those social states to which they are constitutionally entitled, but not more than that. Second, such a system of laws and rules should make a situation possible in which society is stable, that is, in some state of equilibrium. Third, such equilibrium social states should be collectively optimal if possible: there should be no social state that is better for all members of society.

In this monograph, following Gärdenfors (1981), we model rights and constitutions by *effectivity functions*, a term coined by Moulin and Peleg (1982). An effectivity function assigns to each group in society a collection of sets of social states. If a group $S$ is *effective* for a set of social states $B$, then this means that $S$ is constitutionally entitled to the prevailing social state being in $B$. As described above, we additionally need a set of laws and rules in order that $S$ be able to 'enforce' the social state to be in $B$. This is formalized by a *game form*. A game form endows each individual with a set of strategies. To each profile of individual strategies, an outcome function assigns a social state. Thus, a game form makes it possible to impose the constitution in practice. The idea of society functioning as a game is well established. Friedman (1962, p. 25) writes:

> It is important to distinguish the day-to-day activities of people from the general customary and legal framework within which these take place. The day-to-day activities are like the actions of the participants in a game when they are playing it; the framework, like the rules of the game they play. [...]

Naturally, the rules and laws (game form) should reflect the constitution. More precisely, by jointly choosing their individual strategies the members of a group (coalition) $S$ can make sure that the resulting social state (alternative, outcome) is in some set $B$. These combinations $(S, B)$ induced by the game form should be exactly the same as the combinations determined by the constitution. In that case, we say that the game form *represents* the effectivity function which models the constitution.

However, it is not sufficient to have a game form representing the constitution: we also want the game form to possess a certain form of stability. Also this point is stressed by Friedman (1962, p. 25):

> In both games and society also, no set of rules can prevail unless most participants most of the time conform to them without external sanctions; unless that is, there is a broad underlying social consensus. But we cannot rely on custom or on this consensus alone to interpret and to enforce the rules; we need an umpire. These then are the basic roles of government in a free society: to provide a means whereby we can modify the rules, to mediate differences among us on the meaning of the rules, and to enforce compliance with the rules on the part of those few who would otherwise not play the game.

Friedman uses the need for stability (social consensus) as an argument to have an 'umpire' in the form of the government. In our approach, governments do not appear explicitly, but may be represented by a group of individuals in the game form. The same holds for 'law and order': policemen and judges can be players or groups in the game form. This approach does not favor any political points of view: these should be implicit in the constitution or perhaps in the individual preferences of the players. For stability we rely on *game-theoretic equilibrium* concepts. Individuals in society are characterized, not only by their rights and obligations, but also by their preferences. The minimal requirement that we impose is that, whatever preferences the individuals hold, the resulting game (i.e., game form *cum* preferences) has a *Nash equilibrium*.

The issue of representation of a given power structure by a game form can be traced back to von Neumann and Morgenstern (1944), who showed that a superadditive coalitional game can be represented by a (transferable utility) strategic game. This result was extended to games with nontransferable utility, see Aumann (1967) and Borm and Tijs (1992). The most basic theorem in this monograph, Theorem 2.4.7, is in fact the generalization of this result to coalitional game forms[1], that is, effectivity functions. The rest of the monograph originates, conceptually, mainly from work from the mid-seventies by the first author (Peleg, 1978a and 1978b).

The emphasis in this work is on strategic stability of game forms representing effectivity functions, and not so much on considerations of equity or fairness. One could say that these should be embodied in the constitution modelled by the effectivity function and represented by the game form. Nevertheless, just as in the work of Hurwicz and Schmeidler (1978) we shall also pay attention to *Pareto optimality* of equilibrium outcomes of our representing game forms. In fact, it will appear that in the prevalent representing

[1] This term is due to Abdou and Keiding (1991).

game forms constructed in this monograph there are always Pareto optimal equilibrium outcomes.

Summarizing, we study representations of effectivity functions by game forms that satisfy, at least, the minimal stability requirement of having a Nash equilibrium for any profile of individual preferences. Although our leading motivation is to view effectivity functions as modelling constitutions, this is certainly not their exclusive usage. The theory presented here applies also to societies on a smaller scale. In principle, it applies to any society (including, for instance, academic societies) where member rights and obligations are exercised through a set of rules or procedures (e.g., by-laws).

Before we proceed to a more detailed description of the parts and chapters by which this monograph is organized, a few remarks pertaining to both its extent and limitations are in order.

First, from a broader perspective and as mentioned, an effectivity function can be viewed as a general form of a cooperative game, in the spirit of characteristic function games as introduced by von Neumann and Morgenstern (1944). The representation problem is then tantamount to finding a non-cooperative game (form) that endows each player and coalition with the same 'power' as the given effectivity function. The challenge is to find representations that perform well in terms of existence of Nash equilibria (or strong Nash equilibria), and Pareto optimality of the resulting equilibrium outcomes. If possible, the representing game form should be 'nice', e.g., in the sense of some continuity properties. Thus, although constitutions of large and small societies form a leading motivation for this work, the actual scope is larger.

Second, our emphasis is on representation and game-theoretic stability issues, and we do not have the ambition to contribute substantially to the purely legal or philosophical literature on laws and constitutions. Rather, this work can be seen as a specific contribution to the economic literature on *mechanism design* if we take the latter in a wide sense: the design of game forms (mechanisms) in order to reach collective decisions on the basis of individual choices.

Third, the theory in this book should be distinguished from what is usually called *implementation theory*. Implementation theory is concerned with finding a game form associated with a social choice correspondence (or function) such that, for any profile of preferences, the (Nash, strong,...) equilibrium outcomes of the game coincide exactly with the outcomes prescribed by the social choice correspondence. Thus, in implementation theory, one could say that the representation problem is restricted to equilibrium outcomes: there are no restrictions on the outcomes resulting from non-equilibrium behavior. In contrast, in the theory of this book, the representation issue is not restricted to equilibrium outcomes, and, consequently, representing game forms can be constructed completely independently of admissible preference profiles.

This monograph consists of two parts. Part I (Chapters 1–7) is closest to the above general description of the book and considers representations of effectivity functions by game forms, strategic stability properties of those game forms, Pareto optimality of equilibrium outcomes of those game forms, and continuity properties of game forms. Part II (Chapters 8–11) specializes to *social choice functions*. A social choice function assigns an alternative to any profile of preferences and is, thus, a game form where the strategies of the players are their individual preferences. Such a social choice function or, equivalently, such a game form induces an effectivity function which, naturally, is represented by it. Thus, compared to Part I, Part II of the monograph focuses on a special kind of ('direct revelation') game forms, namely social choice functions. The well-known Gibbard-Satterthwaite Theorem says that in such a game form there is always a player who can manipulate, i.e., fares better by not playing the strategy of reporting his true preference. As a consequence, playing the game may lead to undesirable outcomes, and in particular not to the outcomes intended by the original social choice function. For this reason, we shall focus on strong Nash equilibria that result in the same final outcome which would ensue if each player reported his true preference.

The book is based on work that has appeared over the last thirty years, including some recent articles, but also contains new results, such as Nash consistency of upper semicontinuous representations (Chapter 7), and a strongly consistent representation result for topological spaces (Chapter 5). There are quite some new or improved proofs of existing results as well, for instance the proof of the representation theorem in Chapter 2, and many of the proofs in Chapter 3.

## Part I: Representations of constitutions

After the introductory Chapter 1 we set off in Chapter 2 with a formal description of a constitution and the effectivity function that it induces. We follow Gärdenfors (1981) and Peleg (1998): the latter reference presents a more detailed formalization of the concept of rights. The main result of Chapter 2 is Theorem 2.4.7, which shows that every effectivity function (under the usual necessary conditions of monotonicity and superadditivity) can be represented by a game form. The game form used in the proof of this theorem is central: it is used and modified throughout Part I of the monograph.

In Chapter 3 we derive necessary and sufficient conditions on an effectivity function to be representable by a *Nash consistent* game form, i.e., a game form that has a Nash equilibrium for every profile of preferences. This is Theorem 3.2.3. In Theorem 3.3.10 we show that for the case where the number of alternatives (social states) is finite, these conditions can be phrased directly in terms of the original effectivity function. The crucial condition here is an intersection condition that limits the power of individuals. Most of the remainder of this chapter is devoted to the case where the set of alternatives is infinite and we have some topological structure on this set and on the effectivity function. To obtain Nash consistency we need to add a topological condition, but the mentioned intersection condition stays intact.

In this chapter we also discuss the relation between our results and some well-known 'paradoxes' in the social choice literature, notably the Gibbard Paradox (Gibbard, 1974), and the inconsistency of Pareto Optimality and Minimal Liberalism (Sen, 1970), also called the 'liberal paradox'. As to the latter, we show that the game form used to prove the existence of Nash consistent representations – the same game form as used for the main representation result in Chapter 2, see above – is, in fact, *weakly acceptable*. This means that the game formed by the game form and any profile of preferences has a Nash equilibrium with a Pareto optimal outcome. In this sense, our results offer a partial resolution to the liberal paradox.

Chapter 3 is based mainly on Peleg, Peters, and Storcken (2002). The part on the topological case also owes to Abdou (1988).

Chapter 4 goes deeper into the issue of Pareto optimality. Specifically, an *acceptable* game form is a Nash consistent game form such that all Nash equilibrium outcomes for all preference profiles are Pareto optimal. This is clearly desirable: whenever a Nash equilibrium is played, we do not have to worry about its Pareto optimality. Acceptability is attained under the demanding extra condition that no two disjoint coalitions can veto the same alternative $x$: it is not possible that both coalitions $S$ and $T$ are effective for the set of all alternatives except $x$ if $S$ and $T$ are disjoint. Chapter 4 is based mainly on Peleg (2004).

In Chapter 5 we consider representing game forms that are *strongly consistent*, i.e., admit a strong Nash equilibrium for any profile of preferences. A strong Nash equilibrium is a strategy profile that is resistant not only to deviations by individuals but also to deviations of all other coalitions. Strong consistency is a natural strengthening of Nash consistency in view of the fact that we might also expect coalitions to deviate, but it is attained only at the price of strong conditions on the effectivity function, specifically *maximality* and *core stability*. Maximality means that the effectivity function is equal to its polar: for the case of finitely many alternatives, this implies that for each set of alternatives and each coalition of individuals, either that coalition is effective for the given set of alternatives or the complement of that coalition is effective for the complement of the given set of alternatives. Core stability means that for any given profile of preferences there should be an undominated alternative, where an alternative $x$ is undominated if no coalition $S$ is effective for a set $B$ such that all members of $S$ prefer all alternatives of $B$ over $x$. Stability can be replaced by convexity, which is a condition imposed directly on the effectivity function (see Corollary 5.3.4). The results of this chapter are collected from various sources, among which is Peleg (1998).

In Chapter 6 we reexamine the intersection condition necessary for Nash consistency of representing game forms, as established in Chapter 3. Restricting ourselves to the case of finitely many alternatives, we manage to avoid this restrictive condition at the price of allowing some uncertainty in the outcomes. Specifically, we show that adding equal chance lotteries over pure alternatives and assuming that players evaluate these by utility functions respecting stochastic dominance – a minimal and natural requirement – enables us to obtain Nash consistent representations without any extra condition. We call such an effectivity function, obtained by adding equal chance lotteries, a *lottery model* if it preserves the original effectivity function in the following sense: a coalition is effective for a set of lotteries if and only if it was originally effective for the set consisting of the union of the supports of all those lotteries. Chapter 6 may serve as a starting point for finding representations under incomplete information about player types (preferences). It is based on Peleg and Peters (2009).

In the final chapter of Part I we go deeper into the topological properties of representing game forms for the case where the set of alternatives is a compact metric space. Specifically we investigate *continuity* of the outcome function. Our motivation for this is that continuity of the outcome function is a desirable property: it would be undesirable if a small change in an individual strategy would result in an entirely different social state. Unfortunately, we have to start with an impossibility result, entirely due to the lack of continuity of set intersection. On the other hand, we establish some weaker continuity properties, like for instance (upper or lower) semicontinuity when the set of alternatives is a compact subset of the real line. In the analysis the Cantor (ternary) set plays an important role, due to the mathematical fact that there exists a continuous surjective function from the Cantor set to any compact metric space.

The approach in Chapter 7 is necessarily more technical. Nevertheless the main message is that, although completely continuous representations may not exist, there is still much continuity possible while maintaining Nash consistency. This chapter is based mainly on Keiding and Peleg (2006b).

## Part II: Consistent voting

Chapter 8 is introductory to Part II and recalls the Gibbard-Satterthwaite Theorem for social choice functions. As explained above, a social choice function is a special kind of game form, in which the possible preferences of the players are their strategies – so each player reports a strategy – and the outcome function (the social choice function) assigns to each profile of (reported) preferences an alternative. The Gibbard-Satterthwaite Theorem states that there is always a profile of (true) preferences and a player for whom it is not optimal to report his true preference, unless there is a dictator or there are only two alternatives. As a result, the final outcome may not be the desired outcome (desired according to the social choice function applied to the true preference profile). In Part II we accept this state of affairs as a matter of fact and consider the following alternative approach: for a given profile of (true) preferences, can we find a strong Nash equilibrium of the associated game such that the resulting outcome is identical to the outcome we would obtain if each player were to report truthfully? A social choice function with this property is called *exactly and strongly consistent* (ESC). Chapter 8, moreover, very briefly reviews other approaches to escape the consequences of the

Gibbard-Satterthwaite Theorem. One of these is the concept of 'equilibrium with threats' (Peleg and Procaccia, 2007).

Chapter 9 starts the investigation of ESC social choice functions using *feasible elimination procedures*, already treated in Peleg (1978a). The main results of this chapter imply that a(n anonymous) social choice function is ESC if and only if it always selects an alternative that can be obtained by a feasible elimination procedure. Such a feasible elimination procedure is based on a given set of positive integer weights assigned to alternatives. For a given preference profile, an alternative can be eliminated if it is the bottom alternative for a coalition of cardinality at least the weight of the alternative. This is reminiscent of the core of an associated effectivity function and, indeed, a relation between feasible elimination procedures and the core is established (Theorem 9.3.6).

In Chapter 10 the concept of a feasible elimination procedure is extended to apply to effectivity functions. An effectivity function is called *elimination stable* if the set of alternatives resulting from applying feasible elimination procedures is non-empty. The chapter contains characterizations of elimination stable effectivity functions in terms of conditions that can be checked independently of preference profiles, and is based mainly on Holzman (1986b).

In the final chapter (Chapter 11), based on Peleg and Peters (2006), we extend our results to the case where the number of alternatives is (still) finite but the set of individuals is a *continuum*: this is the prevalent framework of (for instance, national) elections. We start by extending the Gibbard-Satterthwaite Theorem to this model and establish, similar to Kirman and Sondermann (1972), that non-manipulability is only possible if there is a so-called 'invisible dictator'. Next, we extend the concept of feasible elimination procedures and find that for subsets of alternatives we have to distinguish between 'e-sets' and 'i-sets': the latter can only be blocked – and hence, eliminated – by coalitions of size strictly larger than the total weight of the 'i-set', whereas for 'e-sets' equality is sufficient for blocking. Using these tools, we are able to extend most of the results of Chapters 9 and 10 to this framework with a continuum of voters. In particular, we obtain an almost complete characterization of anonymous ESC social choice functions in this model.

# Part I
# Representations of constitutions

# Chapter 1
# Introduction to Part I

## 1.1 Motivation and summary

In this chapter we explain why we adopt Gärdenfors's (1981) model of a constitution rather than Arrow's model of a ('well behaved') social welfare function. We start, in Section 1.2, with the definition of a social welfare function and recall some of the questions that it invoked. Then we proceed to formulate Arrow's Impossibility Theorem in Section 1.3. This theorem has severe implications for Arrow's notion of a constitution. We quote Arrow's (1967) account of the dilemma posed by his impossibility theorem.

In Section 1.4 we briefly recall Gärdenfors's definition of a constitution. Then we proceed with a short summary of Part I of this book. We conclude, in Section 1.5, with the remark that within our model a theory is developed which is parallel to Sen's theory of individual rights and liberalism.

## 1.2 Arrow's constitution

As far as we know Arrow was the first to give a precise definition of a constitution in modern social choice theory. We recall the definition. Let $N = \{1, \ldots, n\}$ be the set of members of a society ($n \geq 2$) and let $A = \{a_1, \ldots, a_m\}$ be the set of social states or alternatives ($m \geq 3$). A linear ordering of $A$ is a complete, reflexive, transitive, and antisymmetric binary relation on $A$. For the sake of simplicity we restrict ourselves here to linear orderings, that is, strict preferences. The set of all such linear orderings on $A$ is denoted by $L$. A 'constitution' or *social welfare function* (SWF) is a function $\Phi : L^N \to L$. Thus, a social welfare function associates with

B. Peleg and H. Peters, *Strategic Social Choice*, Studies in Choice and Welfare,
DOI 10.1007/978-3-642-13875-1_1, © Springer-Verlag Berlin Heidelberg 2010

every possible profile of linear orderings, i.e., individual preferences, a social ordering.[1]

In our view at least two questions may be asked about the foregoing definition.

(a) Is it meaningful to regard a social ordering $\Phi(L^N)$ as a preference of the group or society $N$? Some economists argue that only individual preferences are legitimate (see e.g. Buchanan and Tullock, 1962). We think that group preferences may make sense provided one drops the completeness assumption. Indeed, in defining their dominance relation von Neumann and Morgenstern (1944) assume that a coalition may take some action as a result of the coincidence of the preferences of its members over a pair of alternatives. The coalitional preferences in their theory are clearly incomplete. Nevertheless, they lead to interesting solution concepts like stable sets and, later, the core. Furthermore, if completeness is replaced by acyclicity in Arrow's theory then positive results become possible (cf. Mas-Colell and Sonnenschein, 1972).

(b) Is Arrow's notion of a constitution useful as a model for actual, real-life constitutions? In other words, may we deduce conclusions from it that apply to real-life constitutions? Intuitively, the answer to these questions seems to be negative. A constitution is a system of laws that govern the every day life of the members of a society up to interpretation by the judicial authority and modification by the legislative authority. Thus, a constitution is only indirectly related to individual preferences: it does not operate directly on such preferences the way a social welfare function does. Moreover, individual preferences are usually private information, but Arrow's approach ignores game-theoretic aspects (Arrow, 1963, p. 7). As we shall see it is possible to build a game-theoretic model where the definition of a constitution is much closer to the ordinary one. Furthermore, some phenomena that appear in Arrow's model also occur in the game-theoretic model.

## 1.3 Arrow's Impossibility Theorem and its implications

We assume the same model as in the preceding section. For a linear ordering $R^i \in L$ and $x, y \in A$ we also denote $(x, y) \in R^i$ by $x\, R^i\, y$. By $R^i|\{x, y\}$ we denote the restriction of $R^i$ to the set $\{x, y\}$. Similarly, $R^N|\{x, y\}$ denotes the restriction of the profile $R^N \in L^N$ to $\{x, y\}$.

A social welfare function $\Phi$ is *Paretian* if for all $R^N \in L^N$ and all $x, y \in A$ such that $x\, R^i\, y$ for all $i \in N$, we have $x\, \Phi(R^N)\, y$.

A social welfare function $\Phi$ satisfies *independence of irrelevant alternatives* (IIA) if for all $R^N, Q^N \in L^N$ and all $x, y \in A$ such that $R^N|\{x, y\} = Q^N|\{x, y\}$ we have $\Phi(R^N)|\{x, y\} = \Phi(Q^N)|\{x, y\}$.

[1] Strictly speaking, in Arrow (1967) a constitution is not literally an SWF but rather a social choice function derived from an SWF. Nevertheless, the basis is an SWF.

**Theorem 1.3.1 (Arrow's Impossibility Theorem).** *If $\Phi : L^N \to L$ is Paretian and satisfies IIA, then $\Phi$ is dictatorial, that is, there exists $d \in N$ such that $\Phi(R^N) = R^d$ for all $R^N \in L^N$.*

Theorem 1.3.1 has important implications for the possible use of social welfare functions as modelling constitutions. Indeed, Arrow (1967) defined constitutions as 'well behaved' social welfare functions, which means that they should be Paretian and satisfy IIA. But then Arrow's Impossibility Theorem implies that there exists no satisfactory constitution. Arrow's reaction to this is as follows (Arrow, 1967, p. 228):

> This conclusion is quite embarrassing, and it forces us to examine the conditions which have been stated as reasonable. It's hard to imagine anyone quarrelling either with the Pareto Principle or the condition of Non-Dictatorship. The principle of Collective Rationality may indeed be questioned. One might be prepared to allow that the choice from a given environment be dependent on the history of previous choices made in earlier environments, but I think many would find that situation unsatisfactory. There remains, therefore, only the Independence of Irrelevant Alternatives....

Further on (p. 231), he continues:

> Unfortunately, it is clear, as I have already suggested, that social decision processes which are independent of irrelevant alternatives have strong practical advantages, and it remains to be seen whether a satisfactory social decision procedure can really be based on other information.

In order to resolve the described impasse a different definition of a constitution is required. In this book we use Gärdenfors's (1981) definition of a 'rights-system'. Incorporating, as well, Peleg's (1998) representation of a constitution by a game form, we obtain a comprehensive game-theoretic model for social interaction both for large and for small communities. Thus, we also avoid the critique on Arrow's model that it does not take game-theoretic aspects into consideration.

## 1.4 Gärdenfors's model

Gärdenfors (1981) defines a 'rights-system' as follows. A *right* of a non-empty coalition $S \subseteq N$ is a non-empty subset $B \subseteq A$. The interpretation is that $S$ is legally entitled to the containment of the final social alternative in $B$. A *rights-system* $E$ is the set of all pairs $(S, B)$ where $B$ is a right of $S$. Gärdenfors also assumes monotonicity and coherence of rights systems.

This definition of a rights system is formally similar to the definition of an *effectivity function*. We build on this similarity and define a *constitution* as an effectivity function. We also strengthen Gärdenfors's assumption of coherence and assume superadditivity of the constitution. Finally, we use Peleg's (1998) notion of a representation of an effectivity function by a game form to obtain a comprehensive model for the interaction of the members of a social or political system.

In Part I of this book we investigate the following questions. In Chapter 2 we solve the problem of representations of effectivity functions by game forms. A complete characterization of constitutions that may be represented by Nash consistent game forms is given in Chapter 3. Both discrete and continuous models are considered. Chapter 4 characterizes effectivity functions that may be represented by acceptable game forms, a notion that refers to Pareto optimality of Nash equilibria. Representations by strongly consistent game forms are investigated in Chapter 5, where some new results are included compared to the existing literature. Chapter 6 reconsiders representations by Nash-consistent game forms when (even-chance) lotteries on subsets of alternatives are allowed: it is shown that within this framework constitutions have Nash consistent representations without the restrictive assumption established in Chapter 3. Finally, in Chapter 7, we investigate the continuity of the outcome function of a representation when the outcome space and the strategy sets are compact metric spaces.

## 1.5 Notes and comments

The inclusion of the game-theoretic aspects of a political or social system in the definition of a constitution is a deviation from Arrow's definition. Nevertheless, some phenomena appearing in Arrow's theory, which relies on social welfare functions, also appear in our theory, which relies on game forms representing the constitution and on equilibrium notions. A notable example is Sen's (1970) theory of individual rights and liberalism, which is developed within Arrow's model, but has an analog in our model, with results that are different but similar in spirit. See, in particular, Section 3.6.

# Chapter 2
# Constitutions, effectivity functions, and game forms

## 2.1 Motivation and summary

In this chapter we expound on Gärdenfors's (1981) theory of rights-systems or constitutions. Gärdenfors formalizes rights-systems as follows. If $S$ is a coalition (a group of individuals, members, players,...) and $B$ is a set of social states (outcomes, alternatives,...), then $B$ is a 'right' of $S$ in the sense of Gärdenfors if $S$ is legally entitled to the final social state being in $B$. The set of all pairs $(S, B)$ where $S$ is a coalition and $B$ is a right of $S$, is a rights-system. Under very mild conditions a rights-system is a so-called effectivity function. (Effectivity functions are formally introduced in Definition 2.3.1.) Under some additional intuitive conditions, implying the requirements of monotonicity and consistency as postulated in Gärdenfors (1981), a rights-system is a monotonic and superadditive effectivity function (see Section 2.3). Gärdenfors's definition of rights is somewhat indirect, as it is based on attainability of social states. Therefore, we first introduce Peleg's (1998) model of a constitution (Section 2.2). This model distinguishes between rights and social states and describes explicitly how rights of groups result in attainable sets of social states. Nevertheless, for a given assignment of rights the model can be reduced to a rights-system in the sense of Gärdenfors (see Section 2.3). Section 2.2 also contains some important examples – such as the example underlying Gibbard's Paradox (1974) – which are used throughout Part I of this book.

A society cannot function exclusively on the basis of a rights-system or constitution but additionally needs a collection of rules that delimits the actions available and permissible to individuals. For instance, freedom of speech is a basic right in most constitutions but in practice one needs a set of rules that distinguish between an individual's right to express his opinion on the one hand, and slander or discrimination on the other hand. Such rules are the legal means to reach social states that agree with the constitution. As another example, the right to a minimum subsistence level may be part of the

B. Peleg and H. Peters, *Strategic Social Choice*, Studies in Choice and Welfare, DOI 10.1007/978-3-642-13875-1_2, © Springer-Verlag Berlin Heidelberg 2010

constitution, but stealing is usually not regarded as a legal way to satisfy it. Such a collection of rules will be formalized by a game form (Definition 2.4.1). Thus, we search for a game form that 'represents' the effectivity function (constitution). This idea of representation will be given a precise meaning: the game form should endow each group in the society (including, of course, single individuals) with the same possibilities as intended by the constitution. This also implies, basically, that the 'legality' of the game form is judged in terms of the constitution itself. For instance, if stealing is forbidden and, thus, every group of individuals is entitled to a social state where nobody gets robbed, then no individual will have stealing available as a strategy in the game form, since this could lead to a social state where at least one other individual is the victim of theft. In Section 2.4 we prove the existence of a representation for every monotonic and superadditive effectivity function (Theorem 2.4.7). This theorem will be applied and extended throughout Part I.

In Section 2.5 we briefly comment on the possibility of the simultaneous exercising of rights by disjoint coalitions. Section 2.6 concludes with some further remarks.

**Notations.** The following notations are used throughout this book. For an arbitrary set $D$, $|D|$ denotes its cardinality (possibly infinite), $P(D)$ is the collection of all subsets of $D$, and $P_0(D)$ is the collection of all non-empty subsets of $D$. For a subset $C$ of $D$, $C^+$ denotes the collection of all supersets of $C$, that is, $C^+ = \{C' \in P(D) \mid C \subseteq C'\}$.

## 2.2 Constitutions

In this section we present a precise definition of a constitution. Although our definition is based on Gärdenfors (1981), it is more general since we distinguish between rights on the one hand and attainable sets of social states on the other hand.[1] We start with the definition of a society.

**Definition 2.2.1.** A *society* is a list $\mathcal{S} = (N, A, \rho, \alpha, \gamma)$ where

(1) $N$ is the (finite) set of *members* of $\mathcal{S}$.
(2) $A$ is the (finite or infinite) set of *social states*.
(3) $\rho$ is the (finite) set of *rights*.
(4) $\alpha : P(N) \to P(\rho)$, with $\alpha(\emptyset) = \emptyset$, is the (current) *assignment* of rights to groups of members of $\mathcal{S}$.
(5) $\gamma : P(N) \times P(\rho) \to P(P_0(A))$, with $\gamma(\emptyset, \theta) = \emptyset$ and $A \in \gamma(S, \theta)$ for all $\theta \subseteq \rho$ and $S \in P_0(N)$, is the *access correspondence*. Thus, $\gamma$ determines the sets of attainable social states by groups of members of $\mathcal{S}$ as a function of their rights.

[1] A particular consequence is, that this approach allows the change of the constitution as a function of time. See also Remark 2.6.1. However, the possibility of changing constitutions is not further explored in this book.

Some comments on this definition of a society are in order. First, a social state is a complete description of all aspects relevant to the members of society of a possible social situation. Whether the number of social states is finite or infinite depends on the specific application. Sometimes it may be convenient and instructive to model the set of social states as an infinite set, possibly a continuum, with some topological or measure-theoretic structure. Instead of the term 'social state' we often also use the terms *outcome* and *alternative*.

In our definition of a society the set of rights $\rho$ is an abstract set. Intuitively, however, rights are means to reach certain social states. They determine some major aspects of the 'distribution of power' in society $\mathcal{S}$. In our definition, this is reflected by the access correspondence $\gamma$. The definition of this correspondence needs a more detailed explanation. If $S$ is a (non-empty) group of society members and $\theta \subseteq \rho$ is a set of rights, then $\gamma(S,\theta) = \{B_1, \ldots, B_m\}$ is interpreted as follows: if the group $S$ has rights $\theta$, then it is legally entitled to the final social state being in $B_1$ or in $B_2$ or ... or in $B_m$. More precisely, $S$ could insist on the final social state being in $B_1$, or $S$ could insist on the final social state being in $B_2$, etc. But it does not mean that $S$ can insist on the final social state being in the intersection of these sets, assuming that this intersection is non-empty. We shall elaborate on this point in the examples below, and also in Section 2.5, when we discuss the simultaneous exercising of rights by disjoint coalitions. The additional conditions on the access correspondence in (5) above express that the empty coalition is not entitled to anything (this is merely a formal condition) and that any non-empty coalition is at least entitled to the set of all social states. If, in some case, we have $\gamma(S,\theta) = \{A\}$, then this means that $S$ is essentially powerless. In particular, it is usually natural to have $\gamma(S,\emptyset) = \{A\}$ for any (non-empty) coalition $S$.

The first example we consider is a classical example, basically due to Gibbard (1974).

*Example 2.2.2.* Consider a society with two members. Each member has two shirts, a white one and a blue one, and must wear exactly one of the two. Denote $w$ for white and $b$ for blue. Then the set of members is $N = \{1,2\}$ and the set of social states is $A = \{(w,w),(w,b),(b,w),(b,b)\}$, where for each state the first letter refers to the color of 1's shirt and the second letter to the color of 2's shirt. The set of rights is $\rho = \{r_1\}$, where $r_1$ is the right for each member of a group to which $r_1$ is assigned, to choose his own shirt.[2] The rights assignment $\alpha$ is given by $\alpha(\emptyset) = \emptyset$, and $\alpha(1) = \alpha(2) = \alpha(N) = r_1$.[3] The access correspondence $\gamma$ is defined as follows. For all $\theta \subseteq \rho$, $\gamma(\emptyset,\theta) = \emptyset$. Further, $\gamma(S,\emptyset) = \{A\}$ for all non-empty $S \subseteq N$, $\gamma(1,r_1) = \{(w,w),(w,b)\}^+ \cup \{(b,w),(b,b)\}^+$,

[2] Thus, we formalize this right as one and the same right applicable to different groups. Alternatively, it could be modelled as three different rights for the three non-empty groups in this example.

[3] When no confusion is likely we will often denote a singleton $\{a\}$ by $a$.

$\gamma(2, r_1) = \{(w, w), (b, w)\}^+ \cup \{(w, b), (b, b)\}^+$, and $\gamma(N, r_1) = P_0(A)$. We shall return to this example more than once. It has played an important role in the literature, see also Gaertner, Pattanaik, and Suzumura (1992).

We proceed with a somewhat more elaborate example, which is related to another example in Gibbard (1974).

*Example 2.2.3.* Let $N = \{m_1, m_2, f\}$, where $m_i$ is a man, $i = 1, 2$, and $f$ is a woman. We let $A = \{w_1, w_2, s\}$, where $w_i$ is the social state in which $f$ marries $m_i$, $i = 1, 2$, and $s$ denotes the state where $f$ remains single. The set of rights is $\rho = \{r_1, r_2\}$, where $r_1$ is the right to remain single (which is not a vacuous right in some societies), and $r_2$ is the right of a mixed couple to marry (an orthodox society). The assignment of rights is given by $\alpha(m_1) = \alpha(m_2) = \alpha(f) = r_1$ and $\alpha(m_i, f) = r_2$ for $i = 1, 2$. The other groups have no rights as groups. The access correspondence is as follows. If $m_1$ has right $r_1$, then $m_1$ is entitled to the 'final' social state being in the set $\{w_2, s\}$, and, trivially, all supersets: so $\gamma(m_1, r_1) = \{w_2, s\}^+$. By a similar kind of reasoning we have $\gamma(m_1, r_2) = \{A\}$, since for $m_1$ having the right $r_2$ does not give any 'power' (legal entitlement): $m_1$ would need the consent of $f$ to marry her and, moreover, $m_2$ might also have the right to marry $f$, and these rights cannot simultaneously be met if polyandry is prohibited. Table 2.1 presents the complete access correspondence for all non-empty groups.

| rights | $m_1$ | $m_2$ | $f$ | $m_1m_2$ | $m_1f$ | $m_2f$ | $m_1m_2f$ |
|---|---|---|---|---|---|---|---|
| $\emptyset$ | $A$ | $A$ | $A$ | $A$ | $A$ | $A$ | $A$ |
| $r_1$ | $\{s, w_2\}$ | $\{s, w_1\}$ | $\{s\}$ | $\{s\}$ | $\{s\}$ | $\{s\}$ | $\{s\}$ |
| $r_2$ | $A$ | $A$ | $A$ | $A$ | $\{w_1\}$ | $\{w_2\}$ | $\{w_1\}, \{w_2\}$ |
| $\rho$ | $\{s, w_2\}$ | $\{s, w_1\}$ | $\{s\}$ | $\{s\}$ | $\{w_1\}, \{s\}$ | $\{w_2\}, \{s\}$ | $\{x\}$, $x \in A$ |

**Table 2.1** Description of the access correspondence of Example 2.2.3.

The entry in row $\rho$ and column $m_1f$, for instance, must be read as $\gamma(\{m_1, f\}, \rho) = \{w_1\}^+ \cup \{s\}^+$. This means that the group consisting of man $m_1$ and woman $f$ is entitled to the social state where the two get married and also to the social state where both remain single. Further, $\gamma(N, \rho) = \gamma(\{m_1, m_2, f\}, \rho) = P_0(A)$.

The constituents of a society which are directly connected with rights form the constitution.

**Definition 2.2.4.** Let $\mathcal{S} = (N, A, \rho, \alpha, \gamma)$ be a society. The triple $(\rho, \alpha, \gamma)$ is called a *constitution.*

Thus, a constitution consists of a set of rights, an assignment of rights to groups of members of the society, and a function that specifies for each group of members the attainable sets of social states as a function of the rights of the group.

Our definition of a constitution allows for personal rights: there are no *a priori* symmetry conditions imposed on $\alpha$ or $\gamma$. In existing constitutions the same rights are assigned to members of society with similar characteristics. This is of course not ruled out in applications of our definition of a society. For instance, in Example 2.2.3 the two men have the same characteristics and play symmetric roles, which is reflected by both the rights assignment $\alpha$ and the access correspondence $\gamma$.

*Remark 2.2.5.* The above consideration may be formalized by introducing a set of parameters $\pi$ such that each member $i$ of the society is completely specified, for the sake of the analysis of rights and power, by a non-empty subset $\pi_i \subseteq \pi$. Under this assumption, two members $i$ and $j$ can be called *symmetric* if $\pi_i = \pi_j$. Also, the constitution $(\rho, \alpha, \gamma)$ satisfies *equal treatment* if for every pair of symmetric members $i$ and $j$ the transposition $(i, j)$ is a symmetry of the pair $(\alpha, \gamma)$: that is, if $\pi_i = \pi_j$ and $S \subseteq N \setminus \{i, j\}$, then $\alpha(S \cup i) = \alpha(S \cup j)$ and $\gamma(S \cup i, \theta) = \gamma(S \cup j, \theta)$ for every $\theta \subseteq \rho$. For instance, in Example 2.2.3 one could introduce the parameter set $\pi = \{\text{male, female}\}$ and check that indeed $m_1$ and $m_2$ are symmetric.

In a similar vein it is relevant to note that, although individuals may exercise the rights assigned to the groups of which they are members, that does not mean that these rights become individual. For instance, persons over 65 as well as disabled persons may be entitled to free public transportation. In the terminology of the preceding remark, parameters (dummy variables) stating whether a person is over 65 and whether a person is disabled would be included in the overall set of parameters. So a person $i$ may be entitled to free public transportation because he is 65, or because he is disabled, or because he belongs to both groups, but not because he is person $i$. (Of course, it could happen that person $i$ has special exemption from paying public transportation fares, in which case this right is strictly individual.) The same example also shows that there is no contradiction in one group having right $r$ and another overlapping group not having right $r$.

It is also of interest to note that in our model rights should be interpreted in a broad sense: they may include obligations to society, e.g. paying taxes. The observation that a constitution may also contain obligations is not new, see e.g. Kanger and Kanger (1972).

The following example illustrates some of these considerations.

*Example 2.2.6.* Consider a society with $N = M \cup F$, $M \cap F = \emptyset$. The members in $M$ are the males and the members in $F$ the females. For a group $S$, $m_S$ denotes the number of males in $S$ and $f_S$ the number of females. The total number of members is $n$, so $n = m_N + f_N$. There are two rights, $\rho = \{o, r\}$. If a group $S$ has right $o$, then this means that all males in $S$ are obliged to serve in the army. Thus, $o$ is really an obligation. If a group $S$ does not have right (obligation) $o$, then this means that the men in $S$ are not allowed to serve in the army. If a group $S$ has right $r$, then this means that every female in $S$ has the right (but not the obligation) to serve in the army. If a group $S$ does not have right $r$, this means that the women in $S$ are not allowed to serve in the army. (Of course, different interpretations of $o$ and $r$ are possible, and these may lead to different expressions below.)

The set of social states is assumed to be $A = \{0, \ldots, n\}$, where $k \in A$ means that exactly $k$ society members serve in the army.

The access correspondence is as follows. For every non-empty group $S$ we have $\gamma(S, \emptyset) = \{A\}$, and $\gamma(\emptyset, \theta) = \emptyset$ for every $\theta \subseteq \rho$. Further, for any non-empty group $S$:

$$\gamma(S,o) = \{m_S, m_S+1, \ldots, n-f_S\}^+ ,$$
$$\gamma(S,r) = \bigcup_{0\le x\le f_S} \{x, x+1, \ldots, x+n-|S|\}^+ ,$$
$$\gamma(S,\rho) = \bigcup_{0\le x\le f_S} \{x+m_S, x+1+m_S, x+n-f_S\}^+ .$$

The first line reflects the fact that all men in group $S$ have to serve in the army; the fact that the women in $S$ do not have right to serve in the army means that they cannot serve in the army. So $S$ is legally entitled to a social state where the number $k$ of society members who serve in the army is between $m_S$ and $n - f_S$, but cannot decide on the exact value of $k$. In the second expression, the group $S$ can decide how many women $x$ serve in the army; the men in $S$ do not have the obligation to serve, which is interpreted as the impossibility to serve. The third equation reflects the fact that all men in $S$ have to serve and all women in $S$ can choose to serve.

This example illustrates that our definition of a constitution does not *formally* distinguish between rights and obligations. The access correspondence, however, shows that rights and obligations (in this case, $r$ and $o$) play different roles and lead to essential differences in attainable sets of social states.

## 2.3 Constitutions and effectivity functions

Throughout this section let $\mathcal{S} = (N, A, \rho, \alpha, \gamma)$ be a society, with constitution $(\rho, \alpha, \gamma)$. Although the access correspondence $\gamma$ is specified for any assignment of rights, all that matters to determine the actual attainable sets of social states is the assignment of rights $\alpha$. Thus, all the relevant information inherent in the constitution $(\rho, \alpha, \gamma)$ can be summarized by a function $E : P(N) \to P(P_0(A))$ defined by

$$E(S) = E(S; \alpha, \gamma) = \bigcup_{T \subseteq S} \gamma(T, \alpha(T)) \tag{2.1}$$

for every $S \subseteq N$. Hence, according to $E$, group or *coalition* $S$ is entitled to all sets of social states to which some subcoalition of $S$ is legally entitled.

Since, by Definition 2.2.1, $\gamma(\emptyset, \cdot) = \emptyset$, we have $E(\emptyset) = \emptyset$. Also, (2.1) implies that $E$ is *monotonic with respect to coalitions*:[4]

$$S \subseteq T \Rightarrow E(S) \subseteq E(T) \text{ for all } S, T \in P(N). \tag{2.2}$$

Since $A \in \gamma(S, \theta)$ for all $S \in P_0(N)$ and $\theta \subseteq \rho$, we have

$$A \in E(S) \text{ for every } S \neq \emptyset. \tag{2.3}$$

We call the constitution $(\rho, \alpha, \gamma)$ *non-imposed* if the *grand coalition* $N$ has complete power in terms of $E$, that is:

$$E(N) = P_0(A)\,. \tag{2.4}$$

These conditions on the function $E$, except for monotonicity with respect to coalitions, are collected in the concept of an effectivity function. We will formally introduce effectivity functions in a more general framework where not all subsets of $A$ are necessarily admitted as possible sets of social states. More precisely, a *structure* on $A$ is a set $\mathcal{T} \subseteq P_0(A)$ such that (i) $A \in \mathcal{T}$; and (ii) $B_1 \cap B_2 \in \mathcal{T}$ for all $B_1, B_2 \in \mathcal{T}$ with $B_1 \cap B_2 \neq \emptyset$. Examples are situations where $(A, \mathcal{T})$ is a topological or measurable space. Of course, also $\mathcal{T} = P_0(A)$ is a structure. We use the notation $(A, \mathcal{T})$ to refer to a set of alternatives with structure and call this a *structured space*.

**Definition 2.3.1.** For a structured space $(A, \mathcal{T})$, an *effectivity function* (EF) is a function $E : P(N) \to P(\mathcal{T})$ that satisfies (i) $E(\emptyset) = \emptyset$, (ii) (2.3), and (iii) $E(N) = \mathcal{T}$.

Condition (iii) in this definition is equivalent to (2.4) if $\mathcal{T} = P_0(A)$.

The functions $E$ associated according to (2.1) with the examples of Section 2.2 are described in the following example.

*Example 2.3.2.* (i) The function $E$ associated with first Gibbard example concerning the choice of shirt color (Example 2.2.2) is given by $E(\emptyset) = \emptyset$ and:

$$\begin{aligned} E(1) &= \{(w,w),(w,b)\}^+ \cup \{(b,w),(b,b)\}^+\,,\\ E(2) &= \{(w,w),(b,w)\}^+ \cup \{(w,b),(b,b)\}^+\,,\\ E(N) &= P_0(A)\,. \end{aligned}$$

Clearly, $(\rho, \alpha, \gamma)$ is non-imposed and, consequently, $E$ is an effectivity function.

(ii) The function $E$ associated with the second Gibbard example concerning marriage within the trio $\{m_1, m_2, f\}$ (Example 2.2.3) is given by $E(\emptyset) = \emptyset$ and:

$$E(m_1) = \{s, w_2\}^+\,,\ E(m_2) = \{s, w_1\}^+\,,\ E(f) = \{s\}^+\,,$$

[4] This condition is also proposed in Gärdenfors (1981).

$$E(\{m_1, f\}) = \{s\}^+ \cup \{w_1\}^+ \;, E(\{m_2, f\}) = \{s\}^+ \cup \{w_2\}^+ \;,$$

$$E(\{m_1, m_2\}) = \{s, w_2\}^+ \cup \{s, w_1\}^+ \;, \; E(N) = P_0(A) \;.$$

Again, $(\rho, \alpha, \gamma)$ is non-imposed, and consequently, $E$ is an effectivity function.

(iii) For Example 2.2.6 concerning the right or obligation to serve in the army, we assume that every non-empty coalition is assigned $\rho = \{o, r\}$, that is: in every non-empty coalition the men have the obligation to serve in the army while each women has the right to serve in the army. Then $E(\emptyset) = \emptyset$. For every non-empty coalition $S$, every non-empty $T \subseteq S$ and every $0 \leq x \leq f_T$ we have

$$\{x + m_S, \ldots, x + n - f_S\} \subseteq \{x + m_T, \ldots, x + n - f_T\} \;.$$

This implies

$$E(S) = \gamma(S, \rho) = \bigcup_{0 \leq x \leq f_S} \{x + m_S, x + 1 + m_S, x + n - f_S\}^+$$

for every non-empty coalition $S$. In particular,

$$E(N) = \bigcup_{0 \leq x \leq f_N} \{x + m_N\}^+ \subsetneq P_0(A)$$

so $(\rho, \alpha, \gamma)$ violates non-imposition. Of course, this is obvious: since $N$ has obligation $o$, states where not all men serve are not attainable. Thus, $E$ is not an effectivity function. In the present example, if all men in all coalitions are obliged to serve in the army by assumption (rights assignment), then we could reformulate the set of social states by letting these indicate the number of women that serve in the army. Formulated this way, $E$ would become an effectivity function. Alternatively, we can introduce the structure $\mathcal{T}$ consisting of the set $A$ and all sets of social states where all men serve in the army. In that case,

$$E(S) = \bigcup_{0 \leq x \leq f_S} \{x + m_N, x + 1 + m_N, x + n - f_S\}^+, \text{ and}$$

$$E(N) = \bigcup_{0 \leq x \leq f_N} \{x + m_N\}^+ = \mathcal{T},$$

where the superscript '+' should be read as 'all supersets within $\mathcal{T}$'.

Gärdenfors introduced constitutions as effectivity functions. Independently, Moulin and Peleg (1982) introduced effectivity functions in the general context of game theory and especially in its relation to social choice. The concept of an effectivity function can be applied in many directions. For an early summary of the applications of effectivity functions to social choice see Abdou and Keiding (1991).

We continue with further properties of the access correspondence $\gamma$. We say that $\gamma$ is *monotonic with respect to outcomes* if

$$[B \in \gamma(S,\theta),\ B \subseteq B^* \Rightarrow B^* \in \gamma(S,\theta)] \text{ for all } B, B^* \subseteq A,\ S \subseteq N,\ \theta \subseteq \rho. \tag{2.5}$$

Note that in the examples so far this condition is implicit since we always included all supersets in describing the access correspondences. Condition (2.5) simply means that if $S$ can reach the set $B$ of social states or outcomes when $\theta$ is its set of rights, then it can logically reach the larger set $B^*$.

Similarly, we say that $E : P(N) \to P(\mathcal{T})$ is *monotonic with respect to outcomes* if

$$[B \in E(S) \text{ and } B \subseteq B^* \Rightarrow B^* \in E(S)] \text{ for all } B, B^* \in \mathcal{T},\ S \subseteq N. \tag{2.6}$$

*Remark 2.3.3.* A function $E : P(N) \to P(P_0(A))$ that is monotonic with respect to outcomes, i.e. satisfies (2.6), also satisfies (2.4), i.e. $E(N) = P_0(A)$, as soon as $\{a\} \in E(N)$ for all $a \in A$. Thus, under monotonicity with respect to outcomes 'non-imposition' is equivalent to the grand coalition being effective for any single social state.

A function $E : P(N) \to P(\mathcal{T})$ is *monotonic* if it is monotonic both with respect to coalitions (cf. (2.2)) and with respect to outcomes.

A crucial property of a constitution is the property of coherence.[5]

**Definition 2.3.4.** The constitution $(\rho, \alpha, \gamma)$ is *coherent* if for all $S_1, S_2 \in P_0(N)$ with $S_1 \cap S_2 = \emptyset$ and all $B_1 \in \gamma(S_1, \alpha(S_1))$ and $B_2 \in \gamma(S_2, \alpha(S_2))$ we have: $B_1 \cap B_2 \neq \emptyset$.

The intuition behind coherence is straightforward: if, in this definition, $B_1 \cap B_2$ were empty then $S_1$ and $S_2$ could end up in an impossible situation since $S_1$ is entitled to the social state being in $B_1$ whereas $S_2$ is entitled to the social state being in $B_2$. It is easy to see that coherence of the constitution $(\rho, \alpha, \gamma)$ implies and is implied by the analogous condition on the function $E$ defined by (2.1). In what follows, however, we usually need the following stronger condition on an effectivity function.

**Definition 2.3.5.** An effectivity function $E : P(N) \to P(\mathcal{T})$ is *superadditive* if for all $S_1, S_2 \in P_0(N)$ with $S_1 \cap S_2 = \emptyset$ and all $B_1 \in E(S_1)$ and $B_2 \in E(S_2)$ we have: $B_1 \cap B_2 \in E(S_1 \cup S_2)$.

Observe that, if $S \subsetneq T$ and $B \in E(S)$, then superadditivity of the EF $E$ implies $B = B \cap A \in E(S \cup (T \setminus S)) = E(T)$, hence superadditivity implies monotonicity with respect to coalitions.

Also, if $E$ defined by (2.1) is a superadditive EF, then the constitution $(\rho, \alpha, \gamma)$ is coherent. For let $B_1 \in \gamma(S_1, \alpha(S_1))$, $B_2 \in \gamma(S_2, \alpha(S_2))$, and $S_1 \cap S_2 = \emptyset$. Then clearly $B_1 \in E(S_1)$ and $B_2 \in E(S_2)$, so by superadditivity $B_1 \cap B_2 \in E(S_1 \cup S_2)$, which implies $B_1 \cap B_2 \neq \emptyset$.

Gärdenfors (1981) assumes monotonicity with respect to coalitions and coherence for constitutions modelled as effectivity functions.

---

[5] Called 'consistency' in Gärdenfors (1981).

## 2.4 Game forms and a representation theorem

Let $\mathcal{S}$ be a society, and suppose the constitution is described in a concise way by the function $E : P(N) \to P(\mathcal{T})$, as in the preceding section. Recall that $E$ describes for any coalition the sets of social states to which this coalition is legally entitled. It does not tell us how the members of society can actually exercise their rights.

To this end, we assume now that every society member has at its disposal a set of 'legal' strategies. These strategies should be compatible with the constitution in the sense that they endow each coalition with the same legal power as the constitution does. To make this precise, we first introduce the concept of a game form.

**Definition 2.4.1.** A *game form* (GF) is a list $\Gamma = (N; \Sigma^1, \ldots, \Sigma^n; g; A)$ where $N$ is the set of members of the society or *players*; $\Sigma^i$ is the non-empty set of *strategies* of $i \in N$; $g : \Sigma = \Sigma^1 \times \ldots \times \Sigma^n \to A$ is the *outcome function*; and $A$ is the set of social states or *outcomes*.

*Example 2.4.2.* Let $\Gamma = (\{1,2,3\}; \{2,3\}, A, A; g; \{a,b,c\})$, with $g(2,x,y) = x$ and $g(3,x,y) = y$. This is the so-called 'kingmaker' game form: player 1 chooses the king (2 or 3), who in turn chooses an outcome from $A = \{a,b,c\}$. (Cf. Hurwicz and Schmeidler (1978).)

We mentioned that strategies should be 'legal'. We do not give a formal definition of this concept but – informally – call a game form 'legal' if the available strategies do not contradict the assignment of rights. For example, if Adam has the obligation to support his family and stealing is forbidden by law (so by the assignment of rights) then Adam cannot support his family by stealing. That is, stealing is not an available strategy. Moreover, we assume that also coalitions cannot break the law by coordination of their strategies.

We do formalize the preservation of legal power or entitlement resulting from a constitution through a game form by introducing the concept of representation below. First, we associate an effectivity function with each game form.

**Definition 2.4.3.** Let $\Gamma = (N; \Sigma^1, \ldots, \Sigma^n; g; A)$ be a game form, and let $\mathcal{T}$ be a structure on $A$. Let $S \in P_0(N)$, and let $B \in \mathcal{T}$. Then $S$ is *effective* for $B$ if there exists $\sigma_0^S \in \Sigma^S = \prod_{i \in S} \Sigma^i$ such that $g(\sigma_0^S, \sigma^{N \setminus S}) \in B$ for all $\sigma^{N \setminus S} \in \Sigma^{N \setminus S}$. If $g$ is surjective, then $E^\Gamma : P(N) \to P(\mathcal{T})$ defined by $E^\Gamma(\emptyset) = \emptyset$ and

$$E^\Gamma(S) = \{B \in \mathcal{T} \mid S \text{ is effective for } B\} \text{ for all } S \in P_0(N)$$

is the *effectivity function for* $(A, \mathcal{T})$ *associated with* $\Gamma$.

Observe that $E^\Gamma$ is indeed an effectivity function. In particular, surjectivity of $g$ implies that $E^\Gamma(N) = \mathcal{T}$. Effectivity functions associated with game forms were introduced in Moulin and Peleg (1982) as so-called alpha-effectivity functions. It is straightforward to verify that $E^{\Gamma}$ is monotonic and superadditive.

*Example 2.4.4.* For the kingmaker game form $\Gamma$ of Example 2.4.2 the associated effectivity function $E = E^{\Gamma}$ is given by $E(i) = \{A\}$ for each $i \in N$ and $E(S) = P_0(A)$ for $|S| \geq 2$.

*Example 2.4.5.* Let $N = \{1, 2\}$, $A = \{a, b, c, d\}$, and consider the matrix

$$\begin{array}{cc} & \begin{array}{ccc} L & M & R \end{array} \\ \begin{array}{c} T \\ B \end{array} & \begin{pmatrix} a & d & c \\ c & b & d \end{pmatrix} \end{array}$$

where player 1 chooses rows and player 2 columns. This matrix defines a two-person game form $\Gamma = (N; \{T, B\}; \{L, M, R\}; g; A)$ in an obvious way. The associated effectivity function $E^{\Gamma}$ is given by $E^{\Gamma}(1) = \{a, d, c\}^{+} \cup \{c, b, d\}^{+}$, $E^{\Gamma}(2) = \{a, c\}^{+} \cup \{d, b\}^{+} \cup \{c, d\}^{+}$, and $E^{\Gamma}(N) = P_0(A)$. Such a game form is called a *bimatrix* game form since it results in a bimatrix game if utilities of the players on $A$ are added.

The announced idea of representation is one of the main concepts of this part of the book. From now on, let $\mathcal{T}$ be a fixed structure on $A$.

**Definition 2.4.6.** Let $E : P(N) \to P(\mathcal{T})$ be an effectivity function. A game form $\Gamma$ is a *representation* of $E$ if $E^{\Gamma} = E$.

Thus, a game form represents an effectivity function $E$ if its associated effectivity function is equal to $E$. In particular, if $E$ is derived from a constitution as in (2.1), then a representing game form may be considered as a permissible mechanism that enables all the members of the society to exercise their rights simultaneously. Such an effectivity function $E$ may be represented by many different game forms: each of these may be considered as a legal translation of the constitution into strategic behavior. Thus, similar societies and constitutions may be represented quite differently.

Since the effectivity function associated with a game form is monotonic and superadditive, these conditions are necessary for the existence of a representation of an effectivity function. We now show that they are also sufficient.

**Theorem 2.4.7.** *Let $E : P(N) \to P(\mathcal{T})$ be an effectivity function. Then $E$ has a representation if and only if $E$ is monotonic and superadditive.*

*Proof.* For the only-if direction, let $\Gamma = (N; \Sigma^1, \ldots, \Sigma^n; g; A)$ be a representation of $E$. We are done by observing that $E = E^{\Gamma}$ and $E^{\Gamma}$ is monotonic and superadditive.

For the if-direction, assume that $E$ is monotonic and superadditive. We construct a game form that represents $E$. For every $i \in N$ let $N^i = \{S \subseteq N \mid i \in S\}$ and

$$M^i = \{m^i : N^i \to N^i \times \mathcal{T} \mid m_1^i(S) \subseteq S,\ m_2^i(S) \in E(m_1^i(S))\} \tag{2.7}$$

where $m^i(\cdot) = \big(m_1^i(\cdot), m_2^i(\cdot)\big)$ and $m^i$ is monotonic, that is

$$i \in S \subseteq T \Rightarrow m_1^i(S) \subseteq m_1^i(T) \text{ and } m_2^i(T) \subseteq m_2^i(S)\ . \tag{2.8}$$

Observe that $M^i \neq \emptyset$ since it contains the trivial function $S \mapsto (S, A)$. A selection from $\mathcal{T}$ is a function $\varphi : \mathcal{T} \to A$ such that $\varphi(B) \in B$ for every $B \in \mathcal{T}$. Denote by $\Phi$ the set of all selections from $\mathcal{T}$. We define a game form $\Gamma_0 = (N; \Sigma^1, \ldots, \Sigma^n; g_0; A)$ as follows. For each $i \in N$, the set of strategies of $i$ is $\Sigma^i = M^i \times \Phi \times N$. Let $\sigma = (\sigma^1, \ldots, \sigma^n) \in \Sigma^1 \times \ldots \times \Sigma^n$, where $\sigma^i = (m^i, \varphi^i, t^i)$ for every $i \in N$. In order to define $g_0(\sigma)$ we introduce a sequence of partitions of $N$. First, for $S \in P_0(N)$ we define an equivalence relation $\sim_\sigma$ on $S$ by

$$i \sim_\sigma j :\Leftrightarrow m^i(S) = m^j(S), \text{ for all } i, j \in S, \tag{2.9}$$

and denote by $D(S, \sigma)$ the partition of $S$ with respect to $\sim_\sigma$. Now let the first partition of $N$ be $H_0(\sigma) = \{N\}$. If $H_k = \{S_{k,1}, \ldots, S_{k,\ell}\}$ is the $k$-th partition, $k \geq 0$, then we define

$$H_{k+1}(\sigma) = \bigcup_{j=1}^{\ell} D(S_{k,j}, \sigma)\ .$$

Clearly, there exists a minimal $r$ such that $H_k(\sigma) = H_r(\sigma)$ for all $k \geq r$. Write $H_r(\sigma) = \{S_1, \ldots, S_\ell\}$. Then $m_1^i(S_j) = S_j$ and $m_2^i(S_j) = B_j$ for some $B_j \in E(S_j)$, for all $i \in S_j$ and $j = 1, \ldots, \ell$. Since $E$ is superadditive, $B := \bigcap_{j=1}^{\ell} B_j \neq \emptyset$ and $B \in \mathcal{T}$. Let $1 \leq i_0 \leq n$ be the player with $i_0 = (t^1 + \cdots + t^n) \bmod n$. Then we define $g_0(\sigma) = \varphi^{i_0}(B)$.

We prove that $\Gamma_0$ is a representation of $E$. Let $S \in P_0(N)$ and $B \in E(S)$. Choose $\sigma^i = (m^i, \varphi^i, t^i)$ for every $i \in S$ such that $m_1^i(S^*) = S$ and $m_2^i(S^*) = B$ for all $S^* \supseteq S$ and $i \in S$. Then $S$ is an element of the partition $H_r(\sigma^S, \tau^{N \setminus S})$ for each $\tau^{N \setminus S} \in \Sigma^{N \setminus S}$, and $m_2^i(S) = B$ for all $i \in S$. Hence, by definition of $g_0$, $g_0(\sigma^S, \tau^{N \setminus S}) \in B$ for all $\tau^{N \setminus S} \in \Sigma^{N \setminus S}$. So $B \in E^{\Gamma_0}(S)$.

To prove the converse inclusion let $C \in \mathcal{T} \setminus E(S)$. Let $\sigma^S = (m^i, \varphi^i, t^i)_{i \in S} \in \Sigma^S$ and $i_0 \in N \setminus S$ (such an $i_0$ exists since $E(N) = \mathcal{T}$ and thus $S \neq N$). Choose strategies $\tau^i = (m^i, \varphi^i, t^i) \in \Sigma^i$ for every $i \in N \setminus S$, as follows. For every $T \supseteq N \setminus S$ and $i \in N \setminus S$, let $m_1^i(T) = N \setminus S$ and $m_2^i(T) = A$. Further, let $\left(\sum_{i \in S} t^i + \sum_{i \in N \setminus S} t^i\right) \bmod n = i_0$. Clearly, $H_r(\sigma^S, \tau^{N \setminus S}) = \{S_1, \ldots, S_\ell, N \setminus S\}$ for some partition $\{S_1, \ldots, S_\ell\}$ of $S$. Let $B_j = m_2^i(S_j)$, $i \in S_j$, $j = 1, \ldots, \ell$. Then $B_j \in E(S_j)$ for every $j = 1, \ldots, \ell$, so by superadditivity of $E$, $B = \bigcap_{j=1}^{\ell} B_j \in E(S)$. Thus, $B \setminus C \neq \emptyset$ since otherwise, by monotonicity of $E$, $C \in E(S)$, a contradiction. Since $m_2^i(N \setminus S) = A$ for all $i \in N \setminus S$, if $\varphi^{i_0}(B) \in B \setminus C$ then $g_0(\sigma^S, \tau^{N \setminus S}) \notin C$. Hence $C \notin E^{\Gamma_0}(S)$. □

The game form $\Gamma_0$ constructed in the proof of Theorem 2.4.7 is not the unique representation of $E$. Alternative game forms that can be used to prove the theorem can be found in Peleg (1998) and in Peleg, Peters, and Storcken (2002). For reasons that will become clear in the next chapter the game form $\Gamma_0$ will be called a *canonical* representation of $E$.

## 2.5 Representation and simultaneous exercising of rights

Recall from our earlier discussion that a constitution in the sense of Gärdenfors (1981) is simply an effectivity function $E : P(N) \to P(P_0(A))$ – assuming that $P_0(A)$ is the structure on $A$ – and that a right for coalition $S$ is simply any $B \in E(S)$. One may wonder whether rights are non-conflicting or, equivalently, whether they can be exercised simultaneously. For instance, in the first Gibbard (1974) example (Example 2.3.2(i)), the 'rights' $\{(w,w),(w,b)\} \in E(1)$ and $\{(b,w),(b,b)\} \in E(1)$ coincide with the right $\rho_1$ (Example 2.2.2) of player 1 to choose the color of his own shirt. Similarly, player 2 can choose the color of his own shirt or, equivalently, has rights $\{(w,w),(b,w)\}$ and $\{(w,b),(w,b)\}$. Clearly, these rights can be exercised simultaneously: any of the two rights of player 1 has non-empty intersection with any of the two rights of player 2. This follows from the coherence condition of Gärdenfors and, *a forteriori* from superadditivity of $E$. This implies that it also follows from the existence of a representation of $E$, a fact which is also easy to see directly.

**Proposition 2.5.1.** *Let $\Gamma = (N; \Sigma^1, \ldots, \Sigma^n; g; A)$ be a representation of $E$, and let $S_i \in P_0(N)$ and $B_i \in E(S_i)$ for $i = 1, 2$, such that $S_1 \cap S_2 = \emptyset$. Then $B_1 \cap B_2 \neq \emptyset$.*

*Proof.* Let, for $i = 1, 2$, $\sigma^{S_i} \in \Sigma^{S_i}$ satisfy $g(\sigma^{S_i}, \tau^{N\setminus S_i}) \in B_i$ for all $\tau^{N\setminus S_i} \in \Sigma^{N\setminus S_i}$. Then $g(\sigma^{S_1}, \sigma^{S_2}, \tau^{N\setminus(S_1\cup S_2)}) \in B_1 \cap B_2$ for all $\tau^{N\setminus(S_1\cup S_2)} \in \Sigma^{N\setminus(S_1\cup S_2)}$, so $B_1 \cap B_2 \neq \emptyset$. □

Note that the effectivity function of the second Gibbard example (Examples 2.2.3, 2.3.2(ii)) is superadditive and monotonic, and thus has a representation. The same holds for the adapted version of the effectivity function in Examples 2.2.6 and 2.3.2(iii) for the structure described at the end of Example 2.3.2(iii). In particular, in all these examples rights can be exercised simultaneously.

## 2.6 Notes and comments

This chapter is based mainly on Peleg (1998) and Gärdenfors (1981). The proof of Theorem 2.4.7 has benefited from Peleg, Peters, and Storcken (2002).

In the following remark we place the idea of a constitution as introduced in Section 2.2 in a dynamic perspective.

*Remark 2.6.1.* A constitution $(\rho, \alpha, \gamma)$ is at any given point of time a result of a past political process. In a democracy the constitution at a given time represents the status quo of the rights-system and the assignment of rights. Thus, it may be changed by the legislative institutions by procedures such as voting. So, implicitly, in our model rights are politically determined (cf. Sen, 1997). At each time $t$ the members of society have a preference profile $R^N(t)$ that determines the direction of change. Thus, in our framework the problem of choosing the constitution does not arise since the constitution at a given period determines the possible legal constitutions at the next period. In particular, illegal changes such as *coups d'état* are not covered by our model.

Somewhat related to the previous remark is the concept of a *local effectivity function* (Abdou, 1995; Abdou and Keiding, 2003). In a local effectivity function the effectivity of a coalition depends on the current set of social states. Thus, the concept of a local effectivity function generalizes the concept of an effectivity function as used in this monograph.

The next remark concerns the idea of liberalism, a theme that will reoccur in the first part of this book.

*Remark 2.6.2.* Let $E : P(N) \to P(\mathcal{T})$ represent a constitution. $E$ satisfies *liberalism* if each member $i \in N$ can veto some alternative, that is, for each $i \in N$ there exists some $B_i \in E(i) \setminus \{A\}$. For instance, in Example 2.2.3 each member can veto the possibility that he or she gets married and so the associated effectivity function $E$ (Example 2.3.2(ii)) satisfies liberalism. The same is true for Example 2.2.2, where each member can choose the color of his own shirt. The adapted version of the army example (see the last part of Example 2.3.2(iii)) does not satisfy liberalism, since individual men have no say about the number of women in the army. A game form $\Gamma$ satisfies *liberalism* if the associated effectivity function $E^\Gamma$ satisfies liberalism.

Clearly, liberalism implies the existence of non-trivial rights in the sense of Gärdenfors (1981). $E$ satisfies *minimal liberalism* if there are two different individuals $i$ and $j$ with non-trivial rights, i.e., there are $B_i \in E(i) \setminus \{A\}$ and $B_j \in E(j) \setminus \{A\}$. A game form $\Gamma$ satisfies *minimal liberalism* if the associated effectivity function $E^\Gamma$ does. The relationship between liberalism and Pareto optimality, as in Sen's Liberal Paradox (Sen, 1970), will be explored in Chapters 3 and 4.

# Chapter 3
# Nash consistent representations

## 3.1 Motivation and summary

In Chapter 2 we have seen how a constitution of a society or more formally, an effectivity function, and a set of rules that enable the members of the society to exercise their rights simultaneously, i.e., a game form representing the effectivity function, govern the behavior of the members of a (civilized) state. In this chapter we introduce a new element: the preferences of the society members over the social states.[1]

Given a preference profile $R^N$ of the members of society $N$ and a game form $\Gamma$ that represents the effectivity function $E$, the members of $N$ are engaged in an (ordinal) $n$-person game $(\Gamma, R^N)$ in strategic form. Since in such a game the individuals can choose their strategies independently, there is no guarantee that this will lead to a society that is stable, in the sense that no one would like to change his strategy. Therefore, we will impose in this chapter the minimum requirement of Nash consistency: whatever the individual preferences, the resulting game should have at least one Nash equilibrium. The objective of this chapter is to find necessary and sufficient conditions on an effectivity function to have a Nash consistent representation.

We start in Section 3.2 with a general existence theorem, Theorem 3.2.3, which will be used throughout Part I, with suitable modifications. The proof of this theorem heavily relies on our basic representation theorem, Theorem 2.4.7. The main drawback of Theorem 3.2.3 is that in order to check its second condition, (2.3), one has to examine all permissible profiles of preferences. Therefore, in the subsequent sections, Sections 3.3–3.5, we look for direct conditions on an effectivity function for the existence of a Nash consistent representation.

---

[1] The preferences of an individual $i$ can be seen as determined by his set of parameters $\pi_i$ (see Remark 2.2.5). For instance, the preferences of a public servant are determined by both his social role and his selfish interests.

B. Peleg and H. Peters, *Strategic Social Choice*, Studies in Choice and Welfare, DOI 10.1007/978-3-642-13875-1_3, © Springer-Verlag Berlin Heidelberg 2010 

Section 3.3 considers the case where the set of alternatives is finite. It turns out that existence of a Nash consistent representation is guaranteed if the number of individual rights is 'limited'. To quantify this, we introduce the polar effectivity function $E^*$, due to Abdou (1991), and we obtain a concise necessary and sufficient condition (namely (3.6), an intersection condition on polar sets). This condition is not overly strong, in the sense that it is compatible with liberalism (see Remark 2.6.2). On our way to this result we find two interesting properties of the canonical game form $\Gamma_0$ used for Theorem 3.2.3 and already constructed in the proof of Theorem 2.4.7, namely: (i) it contains all Nash equilibrium outcomes attainable through any Nash consistent representation of the underlying effectivity function, and is thus maximal in this respect; (ii) it contains for every profile of preferences a Pareto optimal Nash equilibrium outcome.

Section 3.4 generalizes the results of Section 3.3 to compact Hausdorff topological spaces of alternatives and topological effectivity functions. In the continuous (nondiscrete) case we need an additional condition to obtain Nash consistent representation, namely closedness of the sets $E(N \setminus i)$ in the space of all closed subsets of the topological space $A$.

Following Abdou (1988) we consider in the fifth section topological veto functions. When $A$ is a compact metric space with a probability measure $\mu$ on its Borel subsets we find sharp inequalities on the topological veto function (which is the appropriate way to describe neutral topological effectivity functions on measure spaces) that guarantee the existence of Nash consistent representations.

Some implications of the existence of Pareto optimal Nash equilibrium outcomes for the canonical game form $\Gamma_0$, in particular in relation to Sen's (1970) 'liberal paradox', are collected in Section 3.6. We end with some notes and comments in Section 3.7.

## 3.2 Existence of Nash consistent representations: a general result

Let $A$ be a set of alternatives, finite or infinite, with at least two elements. A *preference* on $A$ is a complete, reflexive and transitive relation on $A$. The set of all preferences on $A$ is denoted by $W = W(A)$. If $R \in W$ and $x, y \in A$, then $xPy$ means $xRy$ and not $yRx$. For $a \in A$ and $R \in W$ we denote by

$$L(a, R) = \{b \in A \mid aRb\} \tag{3.1}$$

the *lower contour set* of $a$ with respect to $R$, i.e., the set of alternatives to which $a$ is weakly preferred. Clearly, $a \in L(a, R)$. Following the usual notation, we denote by $W^S = \{f \mid f : S \to W\}$ the set of mappings from set $S$ to set $W$.

Let $\Gamma = (N; \Sigma^1, \ldots, \Sigma^n; g; A)$ be a game form. An element $R^N \in W^N$ is a *profile* of preferences. For $R^N \in W^N$, the pair $(\Gamma, R^N)$ defines a(n ordinal) game in strategic form. We denote by $\Sigma = \prod_{i \in N} \Sigma^i$ the set of all strategy combinations. A strategy combination $\sigma \in \Sigma$ is a *Nash equilibrium* (NE) of $(\Gamma, R^N)$ if

$$g(\sigma) R^i g(\sigma^{N\setminus\{i\}}, \tau^i) \text{ for all } i \in N \text{ and } \tau^i \in \Sigma^i \,, \tag{3.2}$$

where $\sigma^{N\setminus\{i\}}$ denotes the restriction of $\sigma$ to $N \setminus \{i\}$. The set of all Nash equilibria of the game $(\Gamma, R^N)$ is denoted by $NE(\Gamma, R^N)$.

Let $\mathcal{T}$ be a structure on $A$. We call $\mathcal{T}$ *rich* if $\{a\} \in \mathcal{T}$ for every $a \in A$. For instance, the set of all closed subsets of a Hausdorff topological space (a case that we will consider) is rich.

We call a set of preferences $Q \subseteq W$ *compatible* with $\mathcal{T}$ if for every $R \in Q$ and $a \in A$, $L(a, R) \in \mathcal{T}$. For instance, $P_0(A)$ is a rich structure and every $Q \subseteq W$ is compatible with $P_0(A)$. Other examples will be considered later.

For $Q \subseteq W$, a game form $\Gamma$ is *Nash consistent on* $Q^N$ if $NE(\Gamma, R^N) \neq \emptyset$ for every $R^N \in Q^N$. So a Nash consistent game form has at least one Nash equilibrium for every (permissible) profile of preferences.

The remainder of this section is devoted to proving a general result on the existence of Nash consistent representations of effectivity functions. The proof relies heavily on the representation Theorem 2.4.7. For easy reference we split it up in two propositions.

**Proposition 3.2.1.** *Let $\mathcal{T}$ be a structure and let $Q \subseteq W$ be compatible with $\mathcal{T}$. Let $E : P(N) \to P(\mathcal{T})$ be an effectivity function and let $\Gamma = (N; \Sigma^1, \ldots, \Sigma^n; g; A)$ be a representation of $E$. Let $R^N \in Q^N$, $\sigma \in NE(\Gamma^N, R^N)$, and $x \in A$ with $x = g(\sigma)$. Then $L(x, R^i) \in E(N \setminus i)$ for all $i \in N$.*

*Proof.* Suppose $L(x, R^i) \notin E(N \setminus i)$ for some $i \in N$. Since $\Gamma$ is a representation of $E$ and, thus, $E = E^\Gamma$, we have $L(x, R^i) \notin E^\Gamma(N \setminus i)$. In particular, there must be a $\tau^i \in \Sigma^i$ such that $g(\tau^i, \sigma^{N\setminus i}) \notin L(x, R^i)$, hence $g(\tau^i, \sigma^{N\setminus i}) P^i x$. Since $x = g(\sigma)$, this contradicts the assumption $\sigma \in NE(\Gamma, R^N)$. □

In the following proposition $\Gamma_0$ is the canonical game form for $E$ as constructed in the proof of Theorem 2.4.7.

**Proposition 3.2.2.** *Let $\mathcal{T}$ be a rich structure and let $E : P(N) \to P(\mathcal{T})$ be a monotonic and superadditive effectivity function. Suppose that $R^N \in W^N$ is such that there exists an $x \in A$ with $L(x, R^i) \in E(N \setminus i)$ for all $i \in N$. Then $(\Gamma_0, R^N)$ has a Nash equilibrium $\sigma$ with $g(\sigma) = x$.*

*Proof.* By Theorem 2.4.7 the game form $\Gamma_0$ represents $E$. Choose strategies $\sigma^i = (m^i, \varphi^i, t^i)$ in $\Gamma_0$ for every $i \in N$ such that (i) $m^i(N) = (N, \{x\})$ for every $i \in N$ and (ii) $m^i(N \setminus j) = (N \setminus j, L(x, R^j))$ for all $i \in N \setminus j$, $j = 1, \ldots, n$. (Observe that (i) is possible since $\mathcal{T}$ is rich.) Then, by the definition of $g_0$ (see the proof of Theorem 2.4.7), $g_0(\sigma) = x$ and $g_0(\sigma^{N\setminus\{i\}}, \tau^i) \in L(x, R^i)$ for all $i \in N$ and $\tau^i \in \Sigma^i$. Thus, $\sigma$ is a Nash equilibrium. □

The main result of this section is a direct consequence of Propositions 3.2.1 and 3.2.2 and Theorem 2.4.7.

**Theorem 3.2.3.** *Let $\mathcal{T}$ be a rich structure on $A$ and let $Q \subseteq W$ be compatible with $\mathcal{T}$. Let $E : P(N) \to P(\mathcal{T})$ be an effectivity function. Then $E$ has a representation which is Nash consistent on $Q^N$ if and only if the following two conditions are satisfied:*

(*i*) *$E$ is monotonic and superadditive.*
(*ii*) *For every $R^N \in Q^N$ the following condition holds:*

$$\text{there exists an } x \in A \text{ such that } L(x, R^i) \in E(N \setminus \{i\}) \text{ for all } i \in N. \quad (3.3)$$

## 3.3 The case of finitely many alternatives

Although Theorem 3.2.3 gives a complete and general characterization of effectivity functions for which there is a Nash consistent representation and, in fact, shows that we can take the canonical game form $\Gamma_0$ for such a representation (Proposition 3.2.2), this result is not very well suited for many applications since we have to verify condition (3.3) for all permissible profiles of preferences. In this section we look for direct conditions on effectivity functions to guarantee the existence of Nash consistent representations for the case that $A$ is finite and the structure is the set of all non-empty subsets of $A$. In later sections we consider the same problem for the case where the set of alternatives is infinite but has some topological or measurable structure.

We first need some definitions and results for general effectivity functions. Unless stated otherwise, $(A, \mathcal{T})$ is an arbitrary structured space.

**Definition 3.3.1.** Let $E : P(N) \to P(\mathcal{T})$ be an effectivity function. The *polar* of $E$ is the effectivity function $E^* : P(N) \to P(\mathcal{T})$ defined by $E^*(\emptyset) = \emptyset$ and for $S \in P_0(N)$

$$E^*(S) = \{B \in \mathcal{T} \mid B \cap B' \neq \emptyset \text{ for all } B' \in E(N \setminus S)\}\,. \quad (3.4)$$

Thus, $B \in E^*(S)$ means that $B$ has something in common with *every* set for which the complement $N \setminus S$ is effective. Intuitively, this means that $N \setminus S$ cannot *prevent* $S$ from obtaining an alternative in $B$. The function $E^*$ reflects a weaker effectivity condition than $E$: whereas $E$ tells us what each coalition can guarantee on its own, $E^*$ tells us what each coalition cannot be kept from.[2]

It is easy to verify that, indeed, $E^*$ is an effectivity function: $A \in E^*(S)$ for every $S \in P_0(N)$ and $E^*(N) = \mathcal{T}$ since $E(\emptyset) = \emptyset$. If $B_1 \in E^*(S)$ and $\mathcal{T} \ni B_2 \supseteq B_1$ then obviously $B_2 \cap B' \neq \emptyset$ for all $B' \in E(N \setminus S)$, so $E^*$ is

[2] In the literature this is sometimes called $\beta$-effectivity, as opposed to $\alpha$-effectivity. See Abdou (1991) and Abdou and Keiding (1991).

monotonic with respect to alternatives. Suppose that $E$ is monotonic with respect to coalitions. If $B \in E^*(S)$ and $S \subseteq T$, then for all $B' \in E(N \setminus T)$ we have $B' \in E(N \setminus S)$ and thus $B \cap B' \neq \emptyset$; so $B \in E^*(T)$, and $E^*$ is monotonic with respect to coalitions. In particular, if $E$ is monotonic then $E^*$ is monotonic. (See Section 2.3 for the definitions of these properties.)

The polar of a superadditive effectivity function, however, does not have to be superadditive, as the following example shows.

*Example 3.3.2.* For the polar of Example 2.3.2 concerning the choice of shirt color we have:

$$E^*(1) = \{(w,w),(w,b)\}^+ \cup \{(w,w),(b,b)\}^+ \cup \{(b,w),(w,b)\}^+ \cup \{(b,w),(b,b)\}^+$$

and

$$E^*(2) = \{(w,w),(b,w)\}^+ \cup \{(w,w),(b,b)\}^+ \cup \{(w,b),(b,w)\}^+ \cup \{(w,b),(b,b)\}^+ .$$

Since, for instance, $\{(w,w),(b,b)\} \in E^*(1)$ and $\{(w,b),(b,w)\} \in E^*(2)$, and these sets have empty intersection, $E^*$ is not superadditive.

Let $E : P(N) \to P(\mathcal{T})$ be an effectivity function. We associate with $E$ a function $\widehat{E} : P(N) \to P(\mathcal{T})$ as follows:

$$\widehat{E}(S) = \begin{cases} E^*(i) & \text{if } S = \{i\},\ i \in N \\ \{A\} & \text{if } |S| > 1,\ S \subseteq N \\ \emptyset & \text{if } S = \emptyset\ . \end{cases} \tag{3.5}$$

The function $\widehat{E}$ is called the *residual* of $E$. Our next step is to define the 'core' of the residual, a concept that will be introduced in a more general context later on. Here, we use it for the analysis of Nash consistent representations of $E$. For $i \in N$, $R^i \in W$, $B \in P_0(A)$, and $x \in A \setminus B$, we write $BP^ix$ if $yP^ix$ for all $y \in B$.

**Definition 3.3.3.** Let $E : P(N) \to P(\mathcal{T})$ be an effectivity function and let $R^N \in W^N$. The *core* of $\widehat{E}$ with respect to $R^N$ is the set

$$C(\widehat{E}, R^N) = \{x \in A \mid \text{there are no } i \in N \text{ and } B \in E^*(i) \text{ with } BP^ix\}\ .$$

$\widehat{E}$ is *stable* on $Q^N \subseteq W^N$ if $C(\widehat{E}, R^N) \neq \emptyset$ for every $R^N \in Q^N$.

Hence, if $x$ is an element of the core $C(\widehat{E}, R^N)$, then no individual $i$ can improve upon $x$, that is, has a set of alternatives for which he is $E^*$-effective and which has all elements preferred to $x$.

Non-emptiness of the cores $C(\widehat{E}, R^N)$ is closely connected to condition (3.3) in Theorem 3.2.3, which is a necessary condition for the existence of a Nash consistent presentation. Very roughly, this condition can be interpreted as the sets $E(N \setminus i)$ being 'large', which implies that the polars $E^*(i)$ should be 'small'. In turn this allows the cores $C(\widehat{E}, R^N)$ to be non-empty. To become precise, we have the following lemma.

**Lemma 3.3.4.** *Let $E : P(N) \to P(\mathcal{T})$ be a monotonic effectivity function and let $Q \subseteq W$ be compatible with $\mathcal{T}$. Let $x \in A$ and $R^N \in Q^N$. Then the following two statements are equivalent:*

(*i*) $x \in C(\widehat{E}, R^N)$.
(*ii*) $L(x, R^i) \in E(N \setminus i)$ *for all* $i \in N$.

*Proof.* For the implication $(i) \Rightarrow (ii)$, let $x \in C(\widehat{E}, R^N)$ and $i \in N$. Then $A \setminus L(x, R^i) \notin E^*(i)$. Thus, by definition of $E^*$ there must be some $B' \in E(N \setminus i)$ with $B' \cap (A \setminus L(x, R^i)) = \emptyset$. Hence, $B' \subseteq L(x, R^i)$ and by monotonicity of $E$ and compatibility of $Q^N$, $L(x, R^i) \in E(N \setminus i)$.

For the implication $(ii) \Rightarrow (i)$, let $L(x, R^i) \in E(N \setminus i)$ for all $i \in N$. Then, for all $i \in N$ and $B \in E^*(i)$, we have $B \cap L(x, R^i) \neq \emptyset$. Hence, $x \in C(\widehat{E}, R^N)$. □

Lemma 3.3.4 has the following corollary, which characterizes the existence of Nash consistent representations under a condition for which we still have to check every permissible profile of preferences.

**Corollary 3.3.5.** *Let $\mathcal{T}$ be a rich structure on $A$ and let $Q \subseteq W$ be compatible with $\mathcal{T}$. Let $E : P(N) \to P(\mathcal{T})$ be a monotonic and superadditive effectivity function. Then $E$ has a representation which is Nash consistent on $Q^N$ if and only if $\widehat{E}$ is stable on $Q^N$.*

*Proof.* Straightforward from Theorem 3.2.3 and Lemma 3.3.4. □

Note that, by Proposition 3.2.1 and Lemma 3.3.4, all Nash equilibrium outcomes of the game $(\Gamma, R^N)$ for any representing game form $\Gamma$ are elements of the core $C(\widehat{E}, R^N)$. By Proposition 3.2.2 and Lemma 3.3.4, if $\mathcal{T}$ is rich, it holds that any element of $C(\widehat{E}, R^N)$ is a Nash equilibrium outcome of the particular game $(\Gamma_0, R^N)$. We summarize these facts in another corollary.

**Corollary 3.3.6.** *Let $\mathcal{T}$ be a structure on $A$ and let $Q \subseteq W$ be compatible with $\mathcal{T}$. Let $E : P(N) \to P(\mathcal{T})$ be a monotonic and superadditive effectivity function. Then:*

(*a*) *If $E$ has a Nash consistent representation $\Gamma$ with outcome function $g$, then $g(NE(\Gamma, R^N)) \subseteq C(\widehat{E}, R^N)$ for all $R^N \in Q^N$.*
(*b*) *If $\mathcal{T}$ is rich and $C(\widehat{E}, R^N) \neq \emptyset$ for all $R^N \in Q^N$, then $g_0(NE(\Gamma_0, R^N)) = C(\widehat{E}, R^N)$ for all $R^N \in Q^N$, where $g_0$ is the outcome function of $\Gamma_0$.*

A particular consequence of this corollary is that the canonical game form $\Gamma_0$ admits the maximum number of Nash equilibrium outcomes. That is, for any preference profile any Nash equilibrium outcome in any representing Nash consistent game form is also a Nash equilibrium outcome for that preference profile in the game form $\Gamma_0$. The following example shows that this inclusion can be strict. We first need a definition.

**Definition 3.3.7.** An alternative $x \in A$ is *Pareto optimal* with respect to preference profile $R^N \in W^N$ if for every $y \in A$ there exists $i \in N$ such that $xR^iy$.

*Example 3.3.8.* Consider the kingmaker game form of Example 2.4.2, i.e., $\Gamma = (\{1,2,3\}; \{2,3\}, A, A; g; A)$, where $A = \{a,b,c\}$, with $g(2,x,y) = x$ and $g(3,x,y) = y$, and let $E = E^\Gamma$. Since $E(S) = P_0(A)$ for every coalition $S$ with at least two players, we have $E^*(i) = \{A\}$ for $i = 1,2,3$. Thus, $C(\widehat{E}, R^N) = A$ for every $R^N \in W^N$. Since in every Nash equilibrium of a game $(\Gamma, R^N)$ the chosen alternative is a best alternative of either player 2 or player 3, every Nash equilibrium outcome is Pareto optimal. Obviously, for certain profiles of preferences the set of Pareto optimal outcomes is a strict subset of $A$. The game form $\Gamma$ in this example is not the canonical game form $\Gamma_0$.

Everything so far in this section holds for arbitrary structured spaces $(A, \mathcal{T})$. In the remainder we consider the case of finitely many alternatives and all subsets permissible.

**Proposition 3.3.9.** *Let $2 \leq |A| < \infty$ and let $E : P(N) \to P(P_0(A))$ be a monotonic and superadditive effectivity function. Then $C(\widehat{E}, R^N) \neq \emptyset$ for every $R^N \in W^N$ if and only if*

$$[B_i \in E^*(i) \text{ for all } i \in N] \Rightarrow \bigcap_{i=1}^{n} B_i \neq \emptyset\,. \tag{3.6}$$

*Proof.* For the only-if direction assume, on the contrary, that there exist $B_i' \in E^*(i)$ such that $\bigcap_{i=1}^n B_i' = \emptyset$. Let $R^i \in W$ satisfy $B_i' P^i (A \setminus B_i')$ for every $i \in N$. Then for $x$ to be in $C(\widehat{E}, R^N)$ we would need $x \in B_i'$ for each $i \in N$, hence $x \in \bigcap_{i=1}^n B_i'$, a contradiction.

For the if-part, let $R^N \in W^N$. For each $i \in N$ define

$$B_i = \{x \in A \mid L(x, R^i) \in E(N \setminus i)\}\,.$$

We claim that $B_i \in E^*(i)$. Indeed, let $C \in E(N \setminus i)$. Since $A$ is finite there exists $y \in C$ such that $yR^iz$ for all $z \in C$. By monotonicity of $E$, $L(y, R^i) \in E(N \setminus i)$. So $y \in B_i$ by definition of $B_i$. Hence, $B_i \cap C \neq \emptyset$. Since $C$ was arbitrary, we have $B_i \in E^*(i)$, and this holds for every $i \in N$. By (3.6) there exists $x \in \bigcap_{i=1}^n B_i$. We claim that $x \in C(\widehat{E}, R^N)$. Indeed, for each $i \in N$ let $x_0^i$ be a worst element of $B_i$, i.e., $yR^ix_0^i$ for all $y \in B_i$. Then $xR^ix_0^i$ for each $i$. Clearly, $L(x_0^i, R^i) \in E(N \setminus i)$ for each $i$ by definition of $B_i$. Hence, if $BP^ix$ for some $i$ and $B \subseteq A$ then $B \cap L(x_0^i, R^i) = \emptyset$ and therefore $B \notin E^*(i)$. Thus, $x \in C(\widehat{E}, R^N)$. □

The following theorem follows directly from Corollary 3.3.5 and Proposition 3.3.9.

**Theorem 3.3.10.** *Let $2 \leq |A| < \infty$ and let $E : P(N) \to P(P_0(A))$ be a monotonic and superadditive effectivity function. Then $E$ has a Nash consistent representation on $W^N$ if and only if (3.6) is satisfied.*

This theorem characterizes existence of Nash consistent representations directly in terms of properties of the effectivity function. To see how it can be used we return to the two Gibbard examples.

*Example 3.3.11.* Recall from Example 3.3.2 that, for the Gibbard example concerning the choice of shirt color, we have $\{(w,w),(b,b)\} \in E^*(1)$ and $\{(w,b),\ (b,w)\} \in E^*(2)$, and these sets have empty intersection. So (3.6) is violated, which implies that this effectivity function has no Nash consistent representation. We can also give a clear intuition why this is so. Recall that, in general, $B \in E^*(S)$ means that the complement $N \setminus S$ cannot keep coalition $S$ from achieving an outcome in $B$. Similarly, $\{(w,w),(b,b)\} \in E^*(1)$ means that player 2 cannot keep player 1 from achieving an outcome in $\{(w,w),(b,b)\}$, and $\{(w,b),(b,w)\} \in E^*(2)$ means that player 1 cannot keep player 2 from achieving an outcome in $\{(w,b),(b,w)\}$. Now suppose that player 1 prefers $\{(w,w),(b,b)\}$ over $\{(w,b),(b,w)\}$, and player 2 prefers $\{(w,b),(b,w)\}$ over $\{(w,w),(b,b)\}$. Clearly, this game cannot have a Nash equilibrium: player 1 would 'overrule' any outcome in $\{(w,b),(b,w)\}$ and player 2 any outcome in $\{(w,w),(b,b)\}$.[3]

*Example 3.3.12.* In the 'marriage' example (Example 2.2.3)) we have (see Example 2.3.2(ii)) $E(\{m_2, f\}) = \{s\}^+ \cup \{w_2\}^+$, so that $E^*(m_1) = \{s, w_2\}^+$. Also, $E(\{m_1, f\}) = \{s\}^+ \cup \{w_1\}^+$, so that $E^*(m_2) = \{s, w_1\}^+$. Since $E(\{m_1, m_2\}) = \{s, w_2\}^+ \cup \{s, w_1\}^+$ whereas $E(m_1) = \{s, w_2\}^+$ and $E(m_2) = \{s, w_1\}^+$, it follows that $E$ is not superadditive and therefore has no representation. If we change the assignment of rights in Example 2.2.3 such that $\alpha(\{m_1, m_2\}) = \{r_1\}$, then $E(\{m_1, m_2\}) = \{s\}^+$, so that $E$ becomes not only superadditive but, moreover $E^*(f) = \{s\}^+$ and therefore $s$ is in any intersection as in (3.6). In that case, $E$ has a Nash consistent representation.

We conclude this section by establishing another attractive property of the canonical game form $\Gamma_0$. We have already seen (Corollary 3.3.6) that this game form admits the maximal number of Nash equilibria. The (proof of the) following theorem shows that any outcome that Pareto dominates a Nash equilibrium outcome must itself be a Nash equilibrium outcome. In particular, since $A$ is finite, there always exist Pareto optimal Nash equilibrium outcomes.

**Theorem 3.3.13.** *Let $2 \leq |A| < \infty$ and let $E : P(N) \to P(P_0(A))$ be a monotonic and superadditive effectivity function that has a Nash consistent representation. Then the canonical game form $\Gamma_0$ is a Nash consistent representation of $E$ with the following property: for every $R^N \in W^N$ and every Nash equilibrium outcome $x$ of $(\Gamma_0, R^N)$ there exists a Pareto optimal Nash equilibrium outcome $y$ of $(\Gamma_0, R^N)$ such that $yR^ix$ for every $i \in N$.*

---

[3] An early consideration of Nash consistency of game forms can be found in Gurvich (1989).

*Proof.* Let $\sigma \in NE(\Gamma_0, R^N)$ and $x = g_0(\sigma)$. If $z \in A$ and $zR^i x$ for all $i \in N$, then $L(z, R^i) \supseteq L(x, R^i)$ for each $i$. Since $L(x, R^i) \in E(N \setminus i)$ for all $i$ by Proposition 3.2.1, monotonicity of $E$ implies $L(z, R^i) \in E(N \setminus i)$ for all $i \in N$. Hence, by Proposition 3.2.2, $(\Gamma_0, R^N)$ has a Nash equilibrium with outcome $z$. Since $A$ is finite the set $\{z \in A \mid zR^i x \text{ for all } i \in N\}$ contains a Pareto optimal alternative. □

## 3.4 Nash consistent representations of topological effectivity functions

In this section we drop the assumption that the set of alternatives $A$ is finite. Instead, we assume throughout that $A$ is a topological space and that the structure on $A$ is the collection of all closed sets:

$$\mathcal{K} = \mathcal{K}(A) = \{B \in P_0(A) \mid B \text{ is closed}\} .$$

(Of course, this includes the finite case with structure $P_0(A)$ under the discrete topology.) As before, $N$ is the set of players. An effectivity function $E : P(N) \to P(\mathcal{K})$ – hence with $E(\emptyset) = \emptyset$, $E(N) = \mathcal{K}$, and $A \in E(S)$ for every $S \neq \emptyset$ – is called *topological*. A preference $R \in W$ is *continuous* if for every $a \in A$ the sets $\{b \in A \mid aRb\}$ and $\{b \in A \mid bRa\}$ are closed. By $V$ we denote the set of all continuous preferences on $A$.

The main purpose of this section is to look for necessary and sufficient conditions for the existence of Nash consistent representations of topological effectivity functions on $V^N$. Our first result is the derivation of a necessary condition analogous to the intersection condition (3.6) for finite sets of alternatives, under the assumption that $A$ is a normal topological space. We recall that a topological space is *normal* if for every two disjoint closed sets $B$ and $B'$ there are two disjoint open sets $U$ and $U'$ with $B \subseteq U$ and $B' \subseteq U'$. Before deriving the announced necessary condition (Theorem 3.4.3 below) we need some preliminary topological concepts and results.

A finite family of open subsets $U_1, \ldots, U_m$ of $A$ is a *covering* of $A$ if $\bigcup_{i=1}^m U_i = A$. A finite family of continuous functions $f_i : A \to [0, 1]$, $i = 1, \ldots, m$, such that $\sum_{i=1}^m f_i(x) = 1$ for all $x \in A$, is a *partition of unity*. A partition of unity $f_1, \ldots, f_m$ is *subordinate* to a covering $U_1, \ldots, U_m$ if $f_i(x) = 0$ for all $x \notin U_i$ for $i = 1, \ldots, m$. For a proof of the following lemma see Kelley (1955, p. 171).

**Lemma 3.4.1.** *Let $A$ be a normal space and $U_1, \ldots, U_m$ a covering of $A$. Then there exists a partition of unity $f_1, \ldots, f_m$ which is subordinate to $U_1, \ldots, U_m$.*

This lemma has the following corollary.

**Corollary 3.4.2.** *Let $A$ be a normal space and let $B_1, \ldots, B_m$ be closed subsets of $A$ such that $\bigcap_{i=1}^m B_i = \emptyset$. Then there are open sets $U_1, \ldots, U_m$ such that $U_i \supseteq B_i$ for all $i = 1, \ldots, m$ and $\bigcap_{i=1}^m U_i = \emptyset$.*

*Proof.* Consider the open sets $U'_i = A \setminus B_i$ for each $i = 1, \ldots, m$. Since $\bigcap_{i=1}^m B_i = \emptyset$, these sets are a covering of $A$. By Lemma 3.4.1 there is a partition of unity $f_1, \ldots, f_m$ which is subordinate to $U'_1, \ldots, U'_m$. Now define for each $i = 1, \ldots, m$ the set $U_i = \{x \in A \mid f_i(x) < 1/(2m)\}$. Then $U_i$ is open and contains $B_i$ since $f_i(x) = 0$ for all $x \in B_i$. Moreover, for each $x \in A$, since $\sum_{i=1}^m f_i(x) = 1$, there must be some $i$ such that $x \notin U_i$. So $\bigcap_{i=1}^m U_i = \emptyset$. □

We are now ready to prove the announced necessary condition for the existence of a Nash consistent representation.

**Theorem 3.4.3.** *Let $A$ be a normal space and let the topological effectivity function $E$ have a Nash consistent representation on $V^N$. Then*

$$[B_i \in E^*(i) \textit{ for all } i \in N] \Rightarrow \bigcap_{i=1}^n B_i \neq \emptyset\,. \tag{3.7}$$

*Proof.* Let $\Gamma = (N; \Sigma^1, \ldots, \Sigma^n; g; A)$ be a Nash consistent representation of $E$ on $V^N$. Let $B_i \in E^*(i)$ for every $i \in N$. Assume, contrary to what we want to prove, that $\bigcap_{i=1}^n B_i = \emptyset$. By Corollary 3.4.2 there exist open sets $U_1, \ldots, U_m$ such that $U_i \supseteq B_i$ for all $i = 1, \ldots, m$ and $\bigcap_{i=1}^m U_i = \emptyset$. By Urysohn's Lemma (e.g., Kelley, 1955, p. 115) there are continuous functions $f_i : A \to [0,1]$ $(i \in N)$ which are equal to 1 on $B_i$ and equal to 0 on $A \setminus U_i$. Consider the continuous preferences induced by the functions $f_i$, also denoted by $f_i$, and let $\sigma$ be a Nash equilibrium in the associated game $(\Gamma, (f_i)_{i \in N})$ with outcome $g(\sigma) = x$. Since $\bigcap_{i=1}^m U_i = \emptyset$ there exists $i \in N$ with $x \notin U_i$, so $f_i(x) = 0$. Hence, $L(x, f_i) = \{y \in A \mid f_i(y) \leq 0\} = \{y \in A \mid f_i(y) = 0\}$. Since $\sigma$ is a Nash equilibrium of $(\Gamma, (f_i)_{i \in N})$, by Proposition 3.2.1 we have $L(x, f_i) \in E(N \setminus i)$. As $B_i \in E^*(i)$, $B_i \cap L(x, f_i) \neq \emptyset$. But $f_i$ is equal to 1 on $B_i$ and $f_i$ is equal to 0 on $L(x, f_i)$, a contradiction. □

In contrast to the case of finitely many alternatives (Theorem 3.3.10) the converse of Theorem 3.4.3 is not true, as the following example shows.

*Example 3.4.4.* Let $A = [0,1]$, let $N = \{1,2,3\}$, and let $\lambda$ be the Lebesgue measure on $A$. Consider a topological EF on $A$ satisfying: $E(N) = \mathcal{K}(A)$; $E(S) = \{B \in \mathcal{K}(A) \mid \lambda(B) > \frac{1}{3}\}$, if $|S| = 2$; $E(i) = \{A\}$ for every $i \in N$; and $E(\emptyset) = \emptyset$. Note that $E$ is monotonic and superadditive. In this case, $E^*(i) = \{B \in \mathcal{K} \mid \lambda(B) \geq \frac{2}{3}\}$ for every $i \in N$, and (3.7) is satisfied. Consider the following three functions defined on $A$: $u^1(x) = x$, $u^2(x) = 1 - x$, and $u^3(x) = \max(\frac{1}{2} - x, x - \frac{1}{2})$. Define the profile $R^N \in V^N$ by

$$xR^iy \Leftrightarrow u^i(x) \geq u^i(y), \text{ for all } x, y \in A \text{ and all } i \in N.$$

As may be verified, there is no point $a \in A$ such that $\lambda(L(a, R^i)) > \frac{1}{3}$ for $i = 1, 2, 3$. See Figure 3.1. Hence, Proposition 3.2.1 implies that $E$ has no Nash consistent representation.

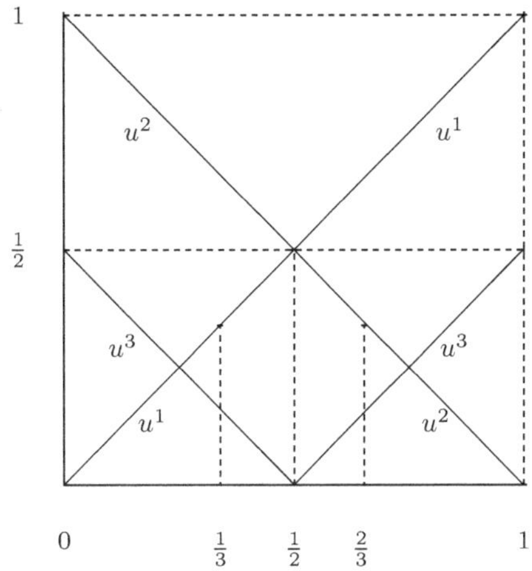


**Fig. 3.1** The utility functions in Example 3.4.4.

In order to obtain sufficient conditions for Nash consistent representation we introduce an additional condition on a topological effectivity function. To this end we need a topology on $\mathcal{K}(A)$.

Let $\top$ denote the topology on $A$. The *upper topology* $\top_u$ is defined by its base

$$\{B \in \mathcal{K}(A) \mid B \subseteq U\},\ U \in \top\ .$$

It follows that the closed sets in $\top_u$ are the intersections of sets of the form

$$\{F \in \mathcal{K}(A) \mid F \cap C \neq \emptyset\},\ C \in \mathcal{K}(A)\ .$$

This implies that, if $E$ is an effectivity function, then the sets $E^*(S)$ are closed in the upper topology for all $S \in P(N)$.

The regularity condition that we will impose is the following:

$$E(N \setminus i) \text{ is closed in } (\mathcal{K}(A), \top_u) \text{ for every } i \in N. \tag{3.8}$$

Note that this condition is not satisfied by the effectivity function in Example 3.4.4, since the sets $\{B \in \mathcal{K}(A) \mid \lambda(B) > \frac{1}{3}\}$ are not closed in the upper topology.[4]

We shall show that (3.7) and (3.8) are sufficient conditions for the existence of a Nash consistent representation for the case where $A$ is a compact Hausdorff space. Recall that a topological space $A$ is Hausdorff if for any two different points $x$ and $y$ there are open disjoint sets $B$ and $C$ with $x \in B$ and

[4] Otherwise, $\{B \in \mathcal{K}(A) \mid \lambda(B) \leq \frac{1}{3}\}$ would be open in the upper topology, but for instance $[0, \frac{1}{3}]$ is only contained in open sets $U$ with $\lambda(U) > \frac{1}{3}$, and such sets contain closed sets of Lebesgue measure larger than $\frac{1}{3}$.

$y \in C$. If $A$ is Hausdorff then, in particular, singletons are closed, so $\mathcal{K}(A)$ is a rich structure.

**Theorem 3.4.5.** *Let $A$ be a compact Hausdorff topological space and let $E : P(N) \to P(\mathcal{K}(A))$ be a monotonic and superadditive effectivity function which satisfies* (3.7) *and* (3.8)*. Then $E$ has a Nash consistent representation on $V^N$.*

*Proof.* Since $V$ is compatible with $\mathcal{K}(A)$ and $\mathcal{K}(A)$ is rich it is sufficient to prove that $E$ satisfies (3.3) in Theorem 3.2.3, i.e., for every $R^N \in V^N$ there exists an $x \in A$ such that $L(x, R^i) \in E(N \setminus \{i\})$ for all $i \in N$.

Let $R^N \in V^N$. For every $i \in N$ let $B_i$ be the set

$$B_i = \{a \in A \mid L(a, R^i) \in E(N \setminus i)\} .$$

Let $F_i = \bigcap\{L(a, R^i) \mid a \in B_i\}$. If $F_i$ is empty then, since all sets $L(a, R^i)$ are closed, the sets $A \setminus L(a, R^i)$ $(a \in B_i)$ are an open covering of $A$. Hence, since $A$ is compact, there is a finite sub-covering $A \setminus L(a_1, R^i), \ldots, A \setminus L(a_m, R^i)$ of $A$ and thus $\bigcap_{j=1}^{m} L(a_j, R^i) = \emptyset$. This is a contradiction since the sets $L(a_j, R^i)$ $(j = 1, \ldots, m)$ are completely ordered by inclusion. Hence, $F_i \neq \emptyset$.

We claim that $F_i \in E(N \setminus i)$. Suppose not then, since $\mathcal{K}(A) \setminus E(N \setminus i)$ is open in the upper topology by (3.8), there is a $U \in \top$ with $F_i \subseteq U$ and for all $B \in \mathcal{K}(A)$ with $B \subseteq U$ we have $B \notin E(N \setminus i)$. In particular, $L(a, R^i) \cap (A \setminus U) \neq \emptyset$ for all $a \in B_i$. By a similar compactness argument as in the preceding paragraph, $\bigcap\{L(a, R^i) \mid a \in B_i\} \cap (A \setminus U) \neq \emptyset$, hence $F_i \cap (A \setminus U) \neq \emptyset$, a contradiction.

By continuity of the preference $R^i$ the compact set $F_i$ has a maximal element $x^i$ with respect to $R^i$. Then $F_i = L(x^i, R^i)$ and by monotonicity of $E$, $B_i = \{a \in A \mid a R^i x^i\}$. In particular, $B_i \in \mathcal{K}(A)$.

We claim that $B_i \in E^*(i)$ for every $i \in N$. Indeed, let $B' \in E(N \setminus i)$ and let $x'$ be a maximal element of $B'$ with respect to $R^i$ (which exists since $B'$ is compact and $R^i$ is continuous). Since $B' \subseteq L(x', R^i)$, monotonicity of $E$ implies $L(x', R^i) \in E(N \setminus i)$. By the definition of $B_i$, it follows that $x' \in B_i \cap B'$. Thus, $B_i \in E^*(i)$.

Since $B_i \in E^*(i)$ for every $i \in N$, (3.7) implies the existence of an $x \in \bigcap_{i=1}^{n} B_i$. Thus, $L(x, R^i) \in E(N \setminus i)$ for every $i \in N$, so that condition (3.3) of Theorem 3.2.3 is satisfied. □

*Remark 3.4.6.* As a representing game form again the canonical game form $\Gamma_0$ constructed in Theorem 2.4.7 can be used. In the above proof Theorem 3.2.3 is used, which in turn is based on Theorem 2.4.7.

In the remainder of this section we extend as much as possible the other results of Section 3.3. The first result is an almost direct consequence of Corollary 3.3.5. It characterizes Nash consistent representation in terms of the residual $\widehat{E}$, defined by (3.5).

**Corollary 3.4.7.** *Let $A$ be a Hausdorff space and let $E : P(N) \to P(\mathcal{K}(A))$ be a monotonic and supperadditive topological effectivity function. Then $E$ has a Nash consistent representation on $V^N$ if and only if $\widehat{E}$ is stable.*

*Proof.* Since a Hausdorff space is rich, i.e., singletons are closed sets, and $V$ is compatible with $\mathcal{K}(A)$, i.e., $L(a, R^i) \in \mathcal{K}(A)$ for all $a \in A$ and $i \in N$ by definition of continuity of preferences, the result follows from Corollary 3.3.5. □

In a similar way the next result follows from Corollary 3.3.6.

**Corollary 3.4.8.** *Let $A$ be a topological space and let $E : P(N) \to P(\mathcal{K}(A))$ be a monotonic and superadditive effectivity function. Then:*

*(a) If $E$ has a Nash consistent representation $\Gamma$ with outcome function $g$, then*

$$g(NE(\Gamma, R^N)) \subseteq C(\widehat{E}, R^N) \text{ for all } R^N \in V^N.$$

*(b) If $A$ is a Hausdorff space and $C(\widehat{E}, R^N) \neq \emptyset$ for all $R^N \in V^N$, then $g_0(NE(\Gamma_0, R^N)) = C(\widehat{E}, R^N)$ for all $R^N \in V^N$, where $g_0$ is the outcome function of $\Gamma_0$.*

Here, $\Gamma_0$ with outcome function $g_0$ is again the game form constructed in the proof of Theorem 2.4.7. Thus also in this case $\Gamma_0$ is a canonical representation of an effectivity function $E : P(N) \to P(\mathcal{K}(A))$ which admits the maximal set of Nash equilibrium outcomes.

We next enquire whether Proposition 3.3.9, which characterizes stability of the residual $\widehat{E}$ in terms of the intersection condition on polar sets $E^*(i)$, can be generalized to topological effectivity functions. Suppose that $A$ is a compact Hausdorff space and that the topological effectivity function $E$ satisfies (3.8), i.e., the sets $E(N \setminus i)$ are closed in the upper topology. Then $E$ has a Nash consistent representation if and only if the intersection condition (3.7) on the polar sets $E^*(i)$ is satisfied. This follows from Theorem 3.4.3 – by observing that a compact Hausdorff space is normal – and Theorem 3.4.5. By combining this observation with Corollary 3.4.7 we obtain the following theorem.

**Theorem 3.4.9.** *Let $A$ be a compact Hausdorff space and let $E : P(N) \to P(\mathcal{K}(A))$ be a monotonic and superadditive effectivity function satisfying (3.8). Then $C(\widehat{E}, R^N) \neq \emptyset$ for all $R^N \in V^N$ if and only if $E$ satisfies (3.7).*

Finally, we consider the question of existence of Pareto optimal Nash equilibrium outcomes. For the finite case we have seen (Theorem 3.3.13) that every Nash equilibrium outcome of a game resulting from using the canonical game form $\Gamma_0$, is (weakly) Pareto dominated by a Pareto optimal Nash equilibrium outcome in the same game. This result generalizes to the case where $A$ is a compact Hausdorff space.

**Theorem 3.4.10.** *Let $A$ be a compact Hausdorff space and let $E : P(N) \to P(\mathcal{K}(A))$ be a monotonic and superadditive effectivity function. For each $R^N \in V^N$, if $x \in g_0(NE(\Gamma_0, R^N))$, then there exists $y \in g_0(NE(\Gamma_0, R^N))$ such that $y$ is Pareto optimal with respect to $R^N$ and $yR^ix$ for all $i \in N$.*

*Proof.* Let $R^N \in V^N$ and $x \in g_0(NE(\Gamma_0, R^N))$, then by Proposition 3.2.1 we have $L(x, R^N) \in E((N \setminus i)$ for every $i \in N$. Let $z \in A$ satisfy $zR^ix$ for every $i \in N$, then by monotonicity of $E$, $L(z, R^N) \in E((N \setminus i)$ for every $i \in N$. By Proposition 3.2.2, $z \in g_0(NE(\Gamma_0, R^N))$. As $A$ is compact, the compact set $\{z \in A \mid zR^ix \text{ for every } i \in N\}$ contains a Pareto optimal alternative $y$. □

We end this section with the following direct consequence of Theorem 3.4.10 (cf. Remark 3.4.6).

**Corollary 3.4.11.** *Under the assumptions of Theorem* 3.4.10*, if in addition $E$ satisfies* (3.7) *and* (3.8)*, then $\Gamma_0$ is a Nash consistent representation of $E$ that has a Pareto optimal Nash equilibrium outcome for every profile of continuous preferences.*

## 3.5 Veto functions

In this section we continue to investigate Nash consistent representation of effectivity functions, but impose the additional condition of 'neutrality': this means that only the *number* of alternatives plays a role in determining effectiveness. This implies that effectivity functions can be described by so-called 'veto functions', saying how many alternatives a coalition can maximally veto. We start with the finite case in Subsection 3.5.1, and consider topological veto functions in Subsection 3.5.2.

### 3.5.1 Finitely many alternatives

Let $A$ be a finite set of alternatives with $|A| = m \geq 2$. An effectivity function $E : P(N) \to P(P_0(A))$ is *neutral* if for every $S \subseteq N$ and $B \in E(S)$, if $B^* \subseteq A$ and $|B^*| = |B|$, then $B^* \in E(S)$.

A *veto function* is a function $v : P(N) \to \{-1, 0, \ldots, m-1\}$ such that $v(\emptyset) = -1$, $v(S) \geq 0$ if $S \neq \emptyset$, and $v(N) = m - 1$. The interpretation of the number $v(S)$ is that the coalition $S$ can veto any subset of alternatives with at most $v(S)$ elements. With a veto function $v$ we can naturally associate an effectivity function $E_v$ by defining

$$E_v(S) = \{B \in P_0(A) \mid \ |A \setminus B| \leq v(S)\} = \{B \in P_0(A) \mid \ |B| \geq m - v(S)\} \tag{3.9}$$

for every $S \in P(N)$. It is straightforward to verify that $E_v$ is indeed an effectivity function. Obviously, $E_v$ is neutral, and $E_v$ is also monotonic with respect to alternatives by definition. Conversely, it is easy to check that for every effectivity function $E$ with these two properties there is a veto function $v$ such that $E = E_v$.

A veto function $v$ is *monotonic* if

$$[S, S^* \in P(N),\ S \subseteq S^*] \Rightarrow v(S) \leq v(S^*) ,$$

and it is *superadditive* if

$$[S, S^* \in P(N),\ S \cap S^* = \emptyset] \Rightarrow v(S) + v(S^*) \leq v(S \cup S^*) .$$

These properties have straightforward interpretations. Monotonicity means that veto power cannot decrease if a coalition increases, and superadditivity means that the union of two disjoint coalitions has at least as much veto power as the two coalitions together have if they act separately. Clearly, a veto function $v$ is monotonic (superadditive) if and only if the associated effectivity function $E_v$ is monotonic (superadditive).

The following example shows that neutral effectivity functions can be induced by simple games.

*Example 3.5.1.* A *simple game* is a pair $(N, \mathbf{W})$ where $\emptyset \neq \mathbf{W} \subseteq P_0(N)$ satisfies $[S \in \mathbf{W}$ and $T \supseteq S] \Rightarrow T \in \mathbf{W}$. The coalitions in $\mathbf{W}$ are called the *winning* coalitions. With a simple game $(N, \mathbf{W})$ we can associate a monotonic and neutral effectivity function $E$ by defining $E(S) = P_0(A)$ if $S \in \mathbf{W}$; $E(S) = \{A\}$ if $S \notin \mathbf{W}$; and $E(\emptyset) = \emptyset$. A simple game $(N, \mathbf{W})$ is *proper* if for each $S \in \mathbf{W}$ we have $N \setminus S \notin \mathbf{W}$. Properness implies in particular that if $S_1 \in \mathbf{W}$ and $S_2 \cap S_1 = \emptyset$, then $S_2 \notin \mathbf{W}$; in turn, it follows that the effectivity function associated with a proper simple game is superadditive.

The existence of Nash consistent representations for monotonic and superadditive neutral effectivity functions can be easily characterized by using Theorem 3.3.10.

**Theorem 3.5.2.** *Let $A$ be a finite set of $m \geq 2$ alternatives and let $v : P(N) \to \{-1, 0, \ldots, m-1\}$ be a monotonic and superadditive veto function. Then the associated effectivity function $E_v$ has a Nash consistent representation if and only if*

$$\sum_{i \in N} v(N \setminus i) > n(m-1) - m . \tag{3.10}$$

*Proof.* For every $i \in N$,

$$E_v^*(i) = \{B \in P_0(A) \mid |B| \geq v(N \setminus i) + 1\} .$$

Hence, (3.6) is satisfied if and only if for all sets $B_1, \ldots, B_n \subseteq A$ with $B_i = v(N \setminus i) + 1$ for each $i \in N$, we have $\bigcap_{i \in N} B_i \neq \emptyset$. This is the case exactly if $\bigcup_{i \in N} A \setminus B_i \neq A$ for all such $B_i$, i.e., if

$$\sum_{i \in N} m - [v(N \setminus i) + 1] < m ,$$

which is the same as (3.10). □

### 3.5.2 Topological veto functions

In order to define neutrality of effectivity functions in case $A$ is an infinite set, we need a measure on $A$. In this section we assume that $A$ is a compact metric space with metric $d$, we let $\mathcal{B}$ denote the $\sigma$-algebra of Borel sets of $A$, and we let $\mu$ be a probability[5] measure on $(A, \mathcal{B})$. This makes it possible to define '$\mu$-neutral' effectivity functions using veto functions.[6]

A *veto function* is now a function $v : P(N) \to [-1, 1]$ with $v(\emptyset) = -1$, $v(N) = 1$, and $v(S) \geq 0$ for all $S \in P_0(N)$. The interpretation is similar as in the finite case: coalition $S$ can veto any (Borel) subset of $A$ of measure at most $v(S)$. The associated effectivity function $E_v$ will be restricted to nonempty closed sets:

$$E_v(S) = \{B \in \mathcal{K}(A) \mid \mu(B) \geq 1 - v(S)\} \text{ for every } S \in P(N).$$

This effectivity function $E_v$ is, indeed, only $\mu$-neutral as $\mu$ may treat different points in $A$ differently. However, if for example $A = [0, 1]$ and $\mu$ is Lebesgue measure, then each $E_v$ is neutral in the ordinary sense.

The definitions of monotonicity and superadditivity of veto functions are identical to the earlier definitions for the finite case and are therefore not repeated. It is easy to check that monotonicity of the veto function implies monotonicity of the associated effectivity function. For superadditivity this is not true, as the following two examples show.

*Example 3.5.3.* Let $A = [0, \frac{1}{2}] \cup [1, \frac{3}{2}]$; $\mu = \lambda$ where $\lambda$ is Lebesgue measure; $N = \{1, 2\}$; and $v : P(N) \to [-1, 1]$ with $v(\emptyset) = \emptyset$, $v(1) = v(2) = \frac{1}{2}$, and $v(N) = 1$. Then $v$ is a monotonic and superadditive veto function. However, $E_v$ is not superadditive: $[0, \frac{1}{2}] \in E_v(1)$, $[1, \frac{3}{2}] \in E_v(2)$, and $[0, \frac{1}{2}] \cap [1, \frac{3}{2}] = \emptyset \notin E_v(N)$.

Observe that the set $A$ in the previous example is not connected. (Recall that a topological space is *connected* if it cannot be written as the union of two nonempty disjoint closed or, equivalently, open sets.) The next example shows however that adding connectedness as a condition on $A$ still does not guarantee superadditivity.

*Example 3.5.4.* Let $A = [0, 2]$; $\mu(B) = \lambda(B \cap [0, 1])$ for every Borel subset $B$ of $A$; $N = \{1, 2\}$; $v(\emptyset) = -1$, $v(1) = 0$, and $v(2) = v(N) = 1$. Then $v$ is

[5] This is without loss of generality: for other nonnegative measures we obtain the same results by adapting the range of the veto function.

[6] The approach in this subsection is motivated by Abdou (1988).

monotonic and superaddditive. Again, $E_v$ is not superadditive: $[0,1] \in E_v(1)$, $[\frac{3}{2},2] \in E_v(2)$, and $[0,1] \cap [\frac{3}{2},2] = \emptyset$.

Observe that the probability measure in the last example does not have full support (see Remark 3.5.5).

*Remark 3.5.5.* A *support* for $\mu$ is any measurable set $B$ such that $\mu(B) = 1$. In our case, where $A$ is a compact metric space and the set of measurable sets is the set of Borel sets, there exists a minimal closed set $B$ with $\mu(B) = 1$, that is, if $C \in \mathcal{K}(A)$ and $\mu(C) = 1$ then $C \supseteq B$ (see Hildenbrand, 1974, p. 49). This set $B$ is called *the support* of $\mu$ and it is denoted by $\mathrm{Supp}(\mu)$. Observe that the requirement $\mathrm{Supp}(\mu) = A$ is equivalent to the requirement that every non-empty open set has positive measure. If this holds, we say that $\mu$ has *full support.*

It turns out that superadditivity is guaranteed if we impose the additional conditions of connectedness and full support.

**Lemma 3.5.6.** *Let $A$ be a connected and compact metric space, let $\mu$ be a probability measure on $(A,\mathcal{B})$ with $Supp(\mu) = A$, and let $v$ be a superadditive veto function. Then $E_v$ is superadditive.*

*Proof.* Let $S_i \in P_0(N)$, $i = 1,2$, $S_1 \cap S_2 = \emptyset$, and $B_i \in E_v(S_i)$, $i = 1,2$. Then $B_i \in \mathcal{K}(A)$ and $\mu(B_i) \geq 1 - v(S_i)$ for $i = 1,2$. Therefore,

$$\begin{aligned}\mu(B_1 \cap B_2) + \mu(B_1 \cup B_2) &= \mu(B_1) + \mu(B_2)\\ &\geq 2 - v(S_1) - v(S_2)\\ &\geq 2 - v(S_1 \cup S_2)\ ,\end{aligned}$$

where the last inequality follows from superadditivity of $v$. Thus,

$$\mu(B_1 \cap B_2) \geq 1 - \mu(B_1 \cup B_2) + 1 - v(S_1 \cup S_2) \geq 1 - v(S_1 \cup S_2)\ . \quad (3.11)$$

This implies superadditivity of $E_v$ if we can show that $B_1 \cap B_2 \neq \emptyset$. Suppose this were not true. Then by (3.11), $\mu(B_1 \cup B_2) = 1$. As $A = \mathrm{Supp}(\mu)$, we must have $B_1 \cup B_2 = A$. But then $A$ is the union of two disjoint nonempty closed sets, contradicting connectedness. □

The first main result in this section is the following theorem, which gives a sufficient condition on a veto function for the associated effectivity function to have a Nash consistent representation.

**Theorem 3.5.7.** *Let $A$ be a connected and compact metric space and let $\mu$ be a probability measure on $(A,\mathcal{B})$ with $Supp(\mu) = A$. Let $v$ be a monotonic and superadditive veto function satisfying*

$$\sum_{i=1}^{n} v(N \setminus i) \geq n - 1\ . \quad (3.12)$$

*Then the associated effectivity function $E_v$ has a Nash consistent representation on $V^N$.*

For the proof of this theorem we need two lemmas. In both lemmas we assume that the conditions of Theorem 3.5.7 are fulfilled. The first lemma characterizes the polar of $E_v$.

**Lemma 3.5.8.** *Let $S \in P_0(N)$. Then*

$$E_v^*(S) = \{B \in \mathcal{K}(A) \mid \mu(B) \geq v(N \setminus S)\}\,.$$

*Proof.* First, suppose that $B \in \mathcal{K}(A)$ with $\mu(B) < v(N \setminus S)$. We show that $B \notin E_v^*(S)$. Define

$$B_t = \{x \in A \mid d(x, B) \geq \frac{1}{t}\}, \quad t = 1, 2, \ldots$$

Then $B_t \in \mathcal{K}(A)$ for every $t = 1, 2, \ldots$ and $\bigcup_{t=1}^{\infty} B_t = A \setminus B$. Hence, there exists a $t_0$ with $\mu(B_{t_0}) \geq 1 - v(N \setminus S)$, which implies $B_{t_0} \in E_v(N \setminus S)$. Since $B_{t_0} \cap B = \emptyset$, it follows that $B \notin E_v^*(S)$.

Next, suppose that $B \in \mathcal{K}(A)$ with $\mu(B) \geq v(N \setminus S)$ and, contrary to what we wish to prove, $B \notin E_v^*(S)$. Then there exists $B' \in E(N \setminus S)$ with $B \cap B' = \emptyset$. As $\mu(B') \geq 1 - v(N \setminus S)$ we must have $\mu(B \cup B') = 1$. Hence, since $\mathrm{Supp}(\mu) = A$, we have $B \cup B' = A$, contradicting connectedness of $A$. □

The next lemma shows that $E_v(\cdot)$ takes values that are closed in the upper topology.

**Lemma 3.5.9.** *For each $S \in P_0(N)$, $E_v(S)$ is closed in the upper topology.*

*Proof.* Let $\alpha \in [0, 1]$. It is sufficient to prove that the set $\mathcal{K}^* = \{B \in \mathcal{K}(A) \mid \mu(B) \geq \alpha\}$ is closed in the upper topology, since each set $E_v(S)$ is of this form. We show that the complement $\mathcal{K}(A) \setminus \mathcal{K}^* = \{B \in \mathcal{K}(A) \mid \mu(B) < \alpha\}$ is open in the upper topology. To show this, it is by definition of the upper topology sufficient to show that if $B_0 \in \mathcal{K}(A) \setminus \mathcal{K}^*$, then there is an open $U \subseteq A$ with $B \subseteq U$ and $\mu(U) < \alpha$. For each $t = 1, 2, \ldots$ define

$$U_t = \{x \in A \mid d(x, B_0) < \frac{1}{t}\}\,.$$

Then $\bigcap_{t=1}^{\infty} U_t = B_0$, hence $\mu(U_t) \to \mu(B_0)$. So we can take $t_0$ with $\mu(U_{t_0}) < \alpha$ and set $U = U_{t_0}$. □

We can now prove Theorem 3.5.7.

*Proof of Theorem 3.5.7.* The EF $E$ is monotonic since $v$ is monotonic, and superadditive by Lemma 3.5.6 since $v$ is superadditive. By Lemma 3.5.9 the set $E_v(N \setminus i)$ is closed in the upper topology for every $i \in N$. In order to apply Theorem 3.4.5 we only have to prove

$$[B_i \in E_v^*(i) \text{ for all } i \in N] \Rightarrow \bigcap_{i=1}^{n} B_i \neq \emptyset\,. \tag{3.13}$$

Assume, on the contrary, that there exist $B_i \in E_v^*(i)$, $i \in N$, such that $\bigcap_{i=1}^{n} B_i = \emptyset$. Let $D_i = A \setminus B_i$, $i \in N$. Clearly, since $\bigcap_{i=1}^{n} B_i = \emptyset$, at least two of the sets $D_i$ are nonempty. Also, each $D_i$ is open, $\bigcup_{i=1}^n D_i = A$, and $\mu(D_i) \leq 1 - v(N \setminus i)$ for each $i \in N$ by Lemma 3.5.8. By (3.12), $\sum_{i=1}^n \mu(D_i) \leq 1$. Suppose that for some $i \neq j$, $D_i \cap D_j \neq \emptyset$. Then $\mu(D_i \cap D_j) > 0$ since $\text{Supp}(\mu) = A$ (cf. Remark 3.5.5). Thus $1 \geq \sum_{k=1}^n \mu(D_k) > \mu(\bigcup_{k=1}^n D_k) = \mu(A) = 1$, a contradiction. Therefore, the sets $D_1, \ldots, D_n$ form a partition of open sets of $A$, at least two of which are nonempty, contradicting connectedness. □

Condition (3.12) in Theorem 3.5.7 is far from necessary for the existence of a Nash consistent representation. Let for instance $A = [0,1]$ and let $\mu$ put weight 9/10 on the one-point set $\{1\}$ and distribute weight 1/10 uniformly over the interval. Consider a monotonic and superadditive veto function $v$ with $v(N \setminus i) = 8/10$ for every $i \in N$. Then for $n > 5$ condition (3.12) is not satisfied but

$$E_v(N \setminus i) = \{B \in \mathcal{K}(A) \mid 1 \in B\} = E_v^*(i)$$

for every $i \in N$. Hence (3.13) is satisfied and Theorem 3.4.5 still implies that $E_v$ has a Nash consistent representation. In this example the singleton $\{1\}$ is an atom of $\mu$. (An *atom* of $\mu$ is a $B \in \mathcal{B}$ such that $\mu(B) > 0$ and for all $B' \in \mathcal{B}$ with $B' \subseteq B$, either $\mu(B') = \mu(B)$ or $\mu(B') = 0$. The measure $\mu$ is *nonatomic* if it has no atoms. For a compact metric space as in our case, $\mu$ is nonatomic precisely if $\mu(x) = 0$ for every $x \in A$.)

In the next theorem we show that for nonatomic measures condition (3.12) is not only sufficient but also necessary for the existence of a Nash consistent representation.

**Theorem 3.5.10.** *Let $A$ be a compact metric space and let $\mu$ be a nonatomic probability measure on $(A, \mathcal{B})$. Let $v$ be a monotonic and superadditive veto function, and let the associated effectivity function $E_v$ have a Nash consistent representation on $V^N$. Then* (3.12) *holds, i.e.*

$$\sum_{i=1}^n v(N \setminus i) \geq n - 1 .$$

*Proof.* Suppose, on the contrary, that

$$\sum_{i=1}^n (1 - v(N \setminus i)) > 1 .$$

Let $N_0 = \{i \in N \mid v(N \setminus i) < 1\}$, hence $N_0 \neq \emptyset$, and choose $0 < \varepsilon < \min\{1 - v(N \setminus i) \mid i \in N_0\}$ such that

$$\sum_{i=1}^n (1 - v(N \setminus i)) > 1 + n\varepsilon . \tag{3.14}$$

Next choose sets $\widehat{B}_i \in \mathcal{B}$, $i \in N$, as follows. If $i \notin N_0$ then $\widehat{B}_i = A$. Since (3.14) implies $\sum_{i \in N_0} [1 - v(N \setminus i) - \varepsilon] > 1$, by nonatomicity of $\mu$ we can choose sets $C_i$ with $\mu(C_i) = 1 - v(N \setminus i) - \varepsilon$ for each $i \in N_0$ and such that

$\bigcup_{i\in N_0} C_i = A$. Then let $\widehat{B}_i = A \setminus C_i$ for each $i \in N_0$. It follows in particular that $\mu(\widehat{B}_i) = v(N \setminus i) + \varepsilon$ for each $i \in N_0$ and $\bigcap_{i\in N_0} \widehat{B}_i = \emptyset$. As $\mu$ is a probability measure on $(A, \mathcal{B})$ we can find closed sets $B_i$, $i \in N$, such that: $B_i = A$ if $i \notin N_0$; $B_i \subseteq \widehat{B}_i$ and $\mu(B_i) > v(N \setminus i)$ for all $i \in N_0$ (see Dunford and Schwartz, 1988, p. 170).

Thus, we have obtained sets $B_i \in E_v^*(i)$, $i \in N$ (cf. Lemma 3.5.8) such that $\bigcap_{i\in N} B_i = \emptyset$. Since a metric space is normal, Theorem 3.4.3 implies that $E_v$ does not have a Nash consistent representation on $V^N$. This is a contradiction, which completes the proof. □

We end this section with an application of Theorems 3.5.7 and 3.5.10.

*Example 3.5.11.* In a city occupying an area of 1 km$^2$ a public facility has to be located. Assume that there are three parties, $N = \{1, 2, 3\}$, and each majority of two parties $\{i, j\}$ can veto any area of at most $0 \leq v(\{i, j\}) \leq 1$. Hence, it is effective for any (closed) area of at least $1 - v(\{i, j\}$ km$^2$. Also, assume that $N$ is effective for any nonempty closed area. Thus, with $v(i) = 0$, $i \in N$, so that single parties are only effective for the whole city, we obtain an effectivity function that is monotonic and superadditive. Theorems 3.5.7 and 3.5.10 imply that it has a Nash consistent representation if and only if

$$v(\{1, 2\}) + v(\{1, 3\}) + v(\{2, 3\}) \geq n - 1 = 2.$$

## 3.6 Liberalism and Pareto optimality of Nash equilibria

In this section we discuss an analogue of Sen's (1970) *Liberal Paradox* within our framework. Sen's result is derived within the classical Arrovian model of a social welfare function, which assigns an ordering of the (finitely many) alternatives to every preference profile. Sen shows that there exists no social decision function (which is a weaker version of a social welfare function) that is Paretian and satisfies minimal liberalism. A social welfare function is Paretian if it orders $x$ above $y$ if every individual does so. It satisfies minimal liberalism if there are at least two individuals each of whom is decisive over some distinct pair of alternatives. An individual is decisive over a pair $x, y$ if the social ordering of $x$ and $y$ coincides with that individual's ordering of $x$ and $y$.

Our notions of liberalism and minimal liberalism are similar to Sen's at least in spirit. Let $A$ be a finite set of social states, let $N = \{1, \ldots, n\}$ be a set of at least two players, and let $E : P(N) \to P(P_0(A))$ be a monotonic and superadditive effectivity function, representing a constitution as in Chapter 2. Recall (see Remark 2.6.2) that $E$ satisfies liberalism if every player can veto some alternative, that is, for every $i \in N$ there exists some $x_i \in A$ such that $A \setminus \{x_i\} \in E(i)$. Note that, since $E$ has a representation (Theorem 2.4.7), individual $i$ can actually enforce $A \setminus \{x_i\}$, and thus veto $x_i$. $E$ satisfies minimal liberalism if there are at least two players who can veto some alternative.

As an analogue of Sen's question in our framework we can ask whether under minimal liberalism Pareto optimality is possible for Nash equilibria of representing game forms. The answer to this question follows easily from our results so far. For instance, the effectivity function of the 'marriage problem' (Example 3.3.12, adapted version with $E(\{m_1, m_2\}) = \{s\}^+$) satisfies liberalism and has a Nash consistent representation $\Gamma$. Thus, by definition (Remark 2.6.2), also $\Gamma$ satisfies liberalism. In particular, we can take $\Gamma = \Gamma_0$, the canonical game form. By Theorem 3.3.13 we know that $\Gamma_0$ has a Pareto optimal Nash equilibrium outcome for every possible profile of preferences. Thus, under this formulation a liberal paradox does not occur. We regard this as a 'partial' resolution (see below) of the liberal paradox and proceed by giving an exact formulation.

The following definition applies for general sets $A$.[7]

**Definition 3.6.1.** Let $\Gamma = (N; \Sigma^1, \ldots, \Sigma^n; g; A)$ be a game form and let $Q$ be some set of preferences. The game form $\Gamma$ is *weakly acceptable* on $Q^N$ if for every $R^N \in Q^N$ there exists $\sigma \in NE(\Gamma, R^N)$ such that $g(\sigma)$ is Pareto optimal with respect to $R^N$.

The same arguments used for the 'marriage problem' example above result in the following corollary on existence of weakly acceptable game forms. Condition (3.6) in this corollary is the familiar intersection condition on polar sets, which by Theorem 3.3.10 implies the existence of a Nash consistent representation. Weak acceptability follows by applying Theorem 3.3.13.

**Corollary 3.6.2.** *Let $A$ be finite. If an effectivity function $E : P(N) \to P(P_0(A))$ is monotonic, superadditive, and satisfies* (3.6), *then $E$ has a weakly acceptable representation on $W^N$.*

The effectivity function $E$ in this corollary may satisfy, in particular, liberalism. Thus, we have within our framework a resolution of the liberal paradox, but this resolution is 'partial' in the sense that in general the canonical game form $\Gamma_0$ also admits Nash equilibrium outcomes that are not Pareto optimal. We go deeper into this issue in Chapter 4.

Results similar to Corollary 3.6.2 can be obtained for topological effectivity functions. Condition (3.6) is replaced by the analogous intersection condition (3.7), and we have to add the closedness condition (3.8). The following result now follows from Corollary 3.4.11.

**Corollary 3.6.3.** *Let $A$ be compact Hausdorff space. If a topological effectivity function $E : P(N) \to P(\mathcal{K}(A))$ is monotonic, superadditive, and satisfies* (3.7) *and* (3.8), *then $E$ has a weakly acceptable representation on the set of all profiles of continuous preferences $V^N$.*

---

[7] Our terminology in this definition is different from that of Hurwicz and Schmeidler (1978).

Thus, we have shown in this section that under suitable conditions there exist weakly acceptable representations for constitutions. The correspondence of Nash equilibrium outcomes of a weakly acceptable game form contains a Pareto optimal alternative for every permissible profile of preferences. This is certainly not typical for arbitrary Nash consistent game forms, as the following example shows.

*Example 3.6.4.* Consider the bimatrix game form

$$\begin{array}{c} \\ T \\ C \\ B \end{array} \begin{array}{c} \begin{array}{ccc} L & M & R \end{array} \\ \begin{pmatrix} b & a & c \\ a & a & a \\ c & a & b \end{pmatrix} \end{array}$$

where player 1 chooses rows and player 2 columns. Clearly, $(C, M)$ is always a Nash equilibrium. If player 1 strictly prefers $b$ to $c$ and $c$ to $a$ and player 2 strictly prefers $c$ to $b$ and $b$ to $a$, then $(C, M)$ is the only Nash equilibrium but $a$ is not Pareto optimal.

## 3.7 Notes and comments

This chapter is based on Peleg, Peters, and Storcken (2002). Compared to the original article there are a few modifications. The main new feature, which was already suggested in the original paper, is the use of the upper topology on the set of nonempty closed subsets of a compact topological space. This simplifies the proofs of some of the results.

*Remark 3.7.1.* Definition 3.3.1 of the polar of an effectivity function is due to Abdou (1991).

*Remark 3.7.2.* An effectivity function $E$ is *maximal* if it is superadditive and $E = E^*$. We shall explain and use this definition in Chapters 4 and 5. A special case of maximality, 1-$(n-1)$ maximality, implies $E(i) = E^*(i)$ for all $i \in N$; with superadditivity this implies our basic intersection condition

$$[B_i \in E^*(i) \text{ for all } i \in N] \Rightarrow \bigcap_{i=1}^{n} B_i \neq \emptyset\,.$$

*Remark 3.7.3.* Let $\Gamma = (N; \Sigma^1, \ldots, \Sigma^n; g; A)$ be a Nash consistent game form on $Q^N$, where $Q \subseteq W$. The game form $\Gamma$ defines a *social choice correspondence* $H : Q^N \to P_0(A)$ by $H(R^N) = g(NE(\Gamma, R^N))$ for every $R^N \in Q^N$. By definition, $\Gamma$ *implements* $H$ in Nash equilibria (see Maskin, 1999). Hence, $H$ is Maskin monotonic on $Q^N$ since this is a necessary condition for implementation (see again Maskin, 1999). A social choice correspondence $H$ is *Maskin monotonic* if the the following condition holds: for all $R^N, \widetilde{R}^N \in Q^N$ and $a \in A$, if $a \in H(R^N)$ and $L(a, R^i) \subseteq L(a, \widetilde{R}^i)$ for all $i \in N$, then $a \in H(\widetilde{R}^N)$.

*Remark 3.7.4.* In Theorem 3.4.5 we can weaken somewhat the assumption that $A$ is a Hausdorff topological space. It is sufficient to require there that $A$ is a $T_1$-space, that is, for every $x \in A$ the singleton $\{x\}$ is closed. This weakening extends to Corollaries 3.4.7 and 3.4.8, but not to Theorem 3.4.9. Hausdorff spaces are also known as $T_2$-spaces. We have chosen to use the Hausdorff property because it is a well known condition.

*Remark 3.7.5.* Condition (3.8) can be weakened: it is sufficient to require that $E(N \setminus i)$ is closed in the *Vietoris topology* $\top_V$ on $\mathcal{K}(A)$, which is finer than the upper topology $\top_u$, i.e., $\top_u \subseteq \top_V$. The Vietoris topology is defined in the following way. The *lower topology* $\top_\ell$ is generated by the sets $\{B \in \mathcal{K}(A) \mid B \cap U \neq \emptyset\}$, $U \in \top$, where $\top$ is the topology on $A$. The Vietoris topology is the common refinement of $\top_u$ and $\top_\ell$. It is sufficient for our use that, if $(A, \top)$ is a compact Hausdorff space, then $(\mathcal{K}(A), \top_V)$ is a compact Hausdorff space as well (Klein and Thompson, 1984, Theorem 2.35 (iii)). In the proof of Theorem 3.4.5, note that the set $\{L(a, R^i) \in \mathcal{K}(A) \mid L(a, R^i) \in E(N \setminus i)\}$ with complete ordering $\supseteq$ is a net, which has $F_i$ as its unique limit point in the compact Hausdorff space $(\mathcal{K}(A), \top_V)$. Since $E(N \setminus i)$ is closed in $(\mathcal{K}(A), \top_V)$, we have $F_i \in E(N \setminus i)$. The rest of the proof is identical to the proof of Theorem 3.4.5.

# Chapter 4
# Acceptable representations

## 4.1 Motivation and summary

In Chapter 3 we have studied the existence of Nash consistent representations of effectivity functions. We have, in fact, shown that the same conditions that guarantee existence of Nash consistent representations also guarantee the existence of weakly acceptable representations, that is, representations that always admit also Pareto optimal Nash equilibria – see Corollaries 3.6.2 and 3.6.3. In this chapter we investigate a subset of the set of Nash consistent game forms, namely the set of acceptable game forms, where a game form is acceptable if: (i) it is Nash consistent and (ii) for every profile of preferences every Nash equilibrium outcome is Pareto optimal. Acceptable game forms were introduced in Hurwicz and Schmeidler (1978). One of the main results of this chapter is a complete characterization of the effectivity functions which can be represented by an acceptable game form. Assuming that the set of social states is a compact Hausdorff topological space and restricting ourselves to continuous preferences we obtain the following result: an effectivity function for at least three players has an acceptable representation if and only if: (i) it has a Nash consistent representation and (ii) no two disjoint coalitions can veto the same alternative. (This follows from Theorem 4.3.1 and Remark 4.6.2. A precise formulation of the latter condition is (4.7).) This result is easy to understand but the proof is quite involved. We outline it here.

First, in Section 4.2, we derive the new necessary condition: if $E$ is the effectivity function of an acceptable game form then no two disjoint coalitions can veto the same alternative (Theorem 4.2.3). In Section 4.3 we formulate the main result (Theorem 4.3.1) and relate it to the results of Chapter 3. Thus, we obtain a neat characterization for finite sets of outcomes (Corollary 4.3.3). Section 4.4 is devoted to the construction of a game form, representing a superadditive and monotonic effectivity function satisfying the new condition (4.7), that has only Nash equilibria with Pareto optimal outcomes. This game form 'extends' the canonical game form $\Gamma_0$ constructed in the proof

B. Peleg and H. Peters, *Strategic Social Choice*, Studies in Choice and Welfare, DOI 10.1007/978-3-642-13875-1_4, © Springer-Verlag Berlin Heidelberg 2010

of Theorem 2.4.7, but it may not be Nash consistent. Finally, in Section 4.5, we complete the proof of Theorem 4.3.1 by further 'extending' the game form of Section 4.4 to an acceptable game form, using some techniques of implementation theory.

The necessary condition (4.7) in conjunction with the familiar intersection conditions for one-person polar sets (3.7) lead to an impossibility theorem: if an effectivity function satisfies minimal liberalism then it has no acceptable representation (Theorem 4.2.4). This result is our version of Sen's Impossibility of a Paretian Liberal. It shows that there is a strong tension between the properties of liberalism and Pareto optimality of Nash equilibrium outcomes of representations. We have seen in Chapter 3 that liberalism of an effectivity function is compatible with the existence of a *weakly* acceptable representation (Corollary 3.6.3). But insisting on acceptable representations even contradicts minimal liberalism.

We also generalize to topological effectivity functions two earlier results on acceptable game forms. First, we extend Theorem 1 in Hurwicz and Schmeidler (1978) for normal spaces (Theorem 4.2.7). Second, again for normal spaces we prove that if an acceptable game form has a maximal effectivity function then this is the effectivity function of a strong simple game – see Corollary 4.2.6. This extends Theorem 3.5 in Dutta (1984). Finally, we prove the existence of an acceptable representation for every effectivity function derived from a proper simple game with at most one vetoer (Proposition 4.6.1).

## 4.2 Acceptable representations and minimal liberalism

Recall from the previous chapter that a game form is *weakly acceptable* if for every profile of admissible preferences there is a Nash equilibrium outcome in the associated game that is Pareto optimal with respect to that profile (Definition 3.6.1). Also recall that the canonical game form $\Gamma_0$, constructed in Chapter 2 and used in Chapter 3 to establish existence of Nash consistent representations of effectivity functions (constitutions) admits a Pareto optimal Nash equilibrium outcome for every profile of preferences and, thus, is weakly acceptable (see Corollaries 3.6.2 and 3.6.3). We have argued that therefore these results offer a 'partial' resolution to Sen's Liberal Paradox. We have used the word 'partial' because in general not every Nash equilibrium of the canonical game form $\Gamma_0$ or, for that matter, some other Nash consistent representing game form has to be Pareto optimal. In other words, finding game forms for which *every* Nash equilibrium outcome is Pareto optimal would be regarded as a 'complete' resolution of the Liberal Paradox in our framework. We shall see in this chapter that this is possible under an additional condition on the effectivity function, namely that no two disjoint coalitions can veto the same alternative. This, however, is quite a strong condition and, in particular, it contradicts minimal liberalism in conjunction with our basic intersection condition necessary for the existence of a Nash consistent representation, condition (3.7).

In order to make these statements precise we first strengthen the weak acceptability concept to 'acceptability' in the following definition. Let $PAR(R^N)$ denote the set of alternatives in $A$ that are Pareto optimal with respect to $R^N$.

**Definition 4.2.1.** A game form $\Gamma = (N; \Sigma^1, \ldots, \Sigma^n; g; A)$ is *acceptable* with respect to $Q \subseteq W$ if the following two conditions are satisfied:

$$NE(\Gamma, R^N) \neq \emptyset \text{ for every } R^N \in Q^N. \tag{4.1}$$

$$g(NE(\Gamma, R^N)) \subseteq PAR(R^N) \text{ for every } R^N \in Q^N. \tag{4.2}$$

Acceptable game forms have been introduced by Hurwicz and Schmeidler (1978). An example of an acceptable game form is the 'kingmaker' game form (Example 2.4.2, see also Example 3.3.8).

We next establish a property of the effectivity function $E^\Gamma$ of an acceptable game form $\Gamma$. This result will be formulated and proved for the case where $A$ is a normal topological space and the structure $\mathcal{K}(A)$ of all nonempty closed sets is rich – thus, every singleton $\{x\}$ $(x \in A)$ is closed[1]. In later sections, we shall need to assume stronger conditions, specifically that $A$ is compact and Hausdorff. Readers (only) interested in the finite case should take notice of the following remark.

*Remark 4.2.2.* The case where $A$ is finite is a special case, with the discrete topology on $A$ (in which all subsets of $A$ are open and hence closed). Any preference on $A$ is continuous in this topology. Alternatively, all results and proofs in this chapter can be read as if $A$ were finite (with $\mathcal{K}(A) = P_0(A)$).

**Theorem 4.2.3.** *Let $A$ be a normal space with rich structure $\mathcal{K}(A)$, and let $\Gamma = (N; \Sigma^1, \ldots, \Sigma^n; g; A)$ be an acceptable game form on $V^N$. Then*

$$\left[S, T \in P_0(N),\ B \in E^\Gamma(S),\ C \in E^\Gamma(T),\ S \cap T = \emptyset\right] \Rightarrow B \cup C = A. \tag{4.3}$$

Note that this condition is equivalent to the requirement that no two disjoint coalitions can veto the same alternative. That is, if $S, T \in P_0(N)$ with $S \cap T = \emptyset$, and $A \setminus \{x\} \in E^\Gamma(S)$ for some $x \in A$, then $A \setminus \{x\} \notin E^\Gamma(T)$. This is straightforward to verify by using the monotonicity of $E^\Gamma$.

*Proof of Theorem 4.2.3.* Suppose that there are $S, T, B, C$ satisfying the premise but not the conclusion of (4.3). Then there exists $x \in A \setminus (B \cup C)$. Since $A$ is normal and $\{x\}$ is closed, by Urysohn's Lemma there exists a continuous function $u : A \to [0, 1]$ such that $u(x) = 1$ and $u(y) = 0$ for all

[1] A normal space with every singleton closed is also called $T_4$-space, cf. Kelley (1955, p. 112).

$y \in B \cup C$.[2] Define now a profile $R^N \in V^N$ by $yR^iz$ if and only if $u(y) \geq u(z)$ for all $y, z \in A$ and $i \in N$. We shall construct a Nash equilibrium with a Pareto non-optimal outcome with respect to $R^N$.

Let $\sigma^S \in \Sigma^S$ satisfy $g(\sigma^S, \tau^{N \setminus S}) \in B$ for all $\tau^{N \setminus S} \in \Sigma^{N \setminus S}$ and let $\sigma^T \in \Sigma^T$ satisfy $g(\sigma^T, \tau^{N \setminus T}) \in C$ for all $\tau^{N \setminus T} \in \Sigma^{N \setminus T}$ (this is possible since $B \in E^\Gamma(S)$ and $C \in E^\Gamma(T)$). Further, let $\sigma^{N \setminus (S \cup T)} \in \Sigma^{N \setminus (S \cup T)}$ be arbitrary. Consider $\sigma = (\sigma^S, \sigma^T, \sigma^{N \setminus (S \cup T)}) \in \Sigma^N$. Then $g(\sigma) \in B \cap C$ and $g(\sigma^{N \setminus i}, \tau^i) \in B \cup C$ for every $i \in N$ and every $\tau^i \in \Sigma^i$. Since $u^i(y) = 0$ for every $i \in N$ and every $y \in B \cup C$, it follows that $\sigma$ is a Nash equilibrium of $(\Gamma, R^N)$. Since $u^i(x) = 1$ for all $i \in N$, it follows that $g(\sigma)$ is Pareto dominated by $x$. □

Recall that an effectivity function $E$ satisfies minimal liberalism if there exist $i, j \in N$, $i \neq j$, and $B_i \in E(i)$, $B_j \in E(j)$ such that $B_i \neq A$ and $B_j \neq A$ (cf. Remark 2.6.2). The first main result of this chapter is, in fact, a negative result.

**Theorem 4.2.4.** *Let $A$ be a normal space with rich structure $\mathcal{K}(A)$, and let the effectivity function $E : P(N) \to P(\mathcal{K}(A))$ satisfy minimal liberalism. Then $E$ has no acceptable representation on $V^N$.*

*Proof.* Suppose, on the contrary, that $E$ has an acceptable representation $\Gamma$, so $E = E^\Gamma$. Let $i, j \in N$, $i \neq j$, $B_i \in E(i)$, $B_j \in E(j)$, and $B_i, B_j \neq A$. By (4.3), $B_i \cup B_j = A$. Thus, we can choose $x \in B_j \setminus B_i$ and $y \in B_i \setminus B_j$. Again by (4.3), $B \cup B_i = A$ for all $B \in E(N \setminus i)$, and thus $x \in B$ for all $B \in E(N \setminus i)$. Similarly, $y \in B$ for all $B \in E(N \setminus j)$. Hence, $\{x\} \in E^*(i)$ and $\{y\} \in E^*(j)$. However, $\{x\} \cap \{y\} = \emptyset$, contradicting the familiar necessary condition for existence of a Nash consistent representation, i.e., (3.7) of Theorem 3.4.3. □

Theorem 4.2.4 is an example of the kind of results that in social choice theory are often referred to as impossibility results. If we insist on minimal liberalism, then the theorem tells us that we have to give up the quest for acceptable representing game forms. However, although minimal liberalism may be an ethically desirable property, it is often not satisfied. Consider for instance a society with set of members $N$ that resolves some class of societal issues by majority rule. If $N$ contains at least three members then this society has no individual rights ($E(i) = \{A\}$ for each $i \in N$), so the condition of minimal liberalism is not satisfied. Nevertheless, it is interesting to know if this 'majority rule society' has an acceptable representation. In somewhat different wording, although minimal liberalism may be compelling as a general principle on a 'macro' social choice level, there may be 'micro' situations – like the example of a group deciding on a collection of relevant issues by majority rule – where minimal liberalism does not apply. In view of these considerations it makes sense to look for acceptable representations and this is, basically, what the rest of this chapter is devoted to.

---

[2] Of course, if $A$ is finite such a function exists trivially.

We shall first characterize acceptable game forms that have a *maximal* effectivity function. Maximality of effectivity functions plays a crucial role in Chapter 5, where we discuss so-called strongly consistent representations. Below, maximal effectivity functions play a role in deriving a result for two-person acceptable game forms.

We recall (see Remark 3.7.2) that an effectivity function $E$ is maximal if it is superadditive and equal to its polar, i.e., $E = E^*$. Observe that superadditivity implies that for every coalition $S$, $E(S) \subseteq \{B \in \mathcal{K}(A) \mid B \cap B' \neq \emptyset$ for all $B' \in E(N \setminus S)\}$, hence $E(S) \subseteq E^*(S)$. Maximality of $E$ implies that these two sets are actually equal. Hence, if $E$ is maximal and $B \notin E(S)$, there must be some $B' \in E(N \setminus S)$ such $B \cap B' = \emptyset$. Since maximality of $E$ also implies monotonicity (see Remark 5.3.1 for an explicit argument), if $A$ is finite then this implies in turn $A \setminus B \in E(N \setminus S)$, which provides another reason for the use of the word 'maximal'. So if $A$ is finite and $E$ is maximal, then for each nonempty subset $B$ of $A$, we have $B \in E(S)$ or $A \setminus B \in E(N \setminus S)$. (If $A$ is not finite, then $A \setminus B$ is not necessarily a closed set.)

We also recall that a simple game is a pair $(N, \mathbf{W})$ where $\emptyset \neq \mathbf{W} \subseteq P_0(N)$ satisfies $[S \in \mathbf{W}$ and $T \supseteq S] \Rightarrow T \in \mathbf{W}$ (Example 3.5.1). A simple game $(N, \mathbf{W})$ is *strong* if for every $S \in P_0(N)$ we have $S \in \mathbf{W} \Leftrightarrow N \setminus S \notin \mathbf{W}$. With a simple game $(N, \mathbf{W})$ we have associated an effectivity function $E$ by letting, for each nonempty coalition $S$, $E(S) = \mathcal{K}(A)$ if $S \in \mathbf{W}$ (i.e., $S$ is winning) and $E(S) = \{A\}$ if $S \notin \mathbf{W}$ (i.e., $S$ is losing). (See again Example 3.5.1.) It is not difficult to verify that the effectivity function associated with a strong simple game is maximal. As to the converse, we have the following relation between maximal effectivity functions and strong simple games. Condition (4.4) in the next theorem is identical to condition (4.3), which we have established as a condition necessary for the existence of a representing acceptable game form.

**Theorem 4.2.5.** *Let $A$ be a normal space with rich structure $\mathcal{K}(A)$ and let $E : P(N) \to P(\mathcal{K}(A))$ be a maximal effectivity function. Then $E$ satisfies*

$$[S_i \in P_0(N),\ B_i \in E(S_i),\ i = 1, 2,\ S_1 \cap S_2 = \emptyset] \Rightarrow B_1 \cup B_2 = A \qquad (4.4)$$

*if and only if $E$ is the effectivity function associated with a strong simple game.*

*Proof.* For the only-if part, let $E$ be a maximal effectivity function satisfying (4.4). Further, let $S \in P_0(N)$, $S \neq N$, and let $x \in A$. We claim that

$$\{x\} \in E(S) \cup E(N \setminus S)\,. \qquad (4.5)$$

Suppose that (4.5) were not true. Then $\{x\} \notin E^*(S)$. Hence, there exists $B_2 \in E(N \setminus S)$ such that $x \notin B_2$. Similarly, there exists $B_1 \in E(S)$ such that $x \notin B_1$. So $B_1 \cup B_2 \subseteq A \setminus \{x\}$, which contradicts (4.4). This proves (4.5). Suppose $\{x\} \in E(S)$. Let $y \in A \setminus \{x\}$. Then, if $\{y\} \notin E(S)$, by (4.5) $\{y\} \in E(N \setminus S)$, hence by superadditivity $\{x\} \cap \{y\} \neq \emptyset$, a contradiction.

Thus, $\{y\} \in E(S)$ for all $y \in A$. As $E$ is monotonic (Remark 5.3.1) this implies $E(S) = \mathcal{K}(A)$. Thus, by (4.4), $E(N \setminus S) = \{A\}$.

We have proved that, for every $S \in P_0(N)$, $S \neq N$, either $E(S) = \mathcal{K}(A)$ or $E(N \setminus S) = \mathcal{K}(A)$. Since $E$ is monotonic, it is the effectivity function associated with a strong simple game.

The proof of the if-part is straightforward. □

An immediate consequence of Theorems 4.2.3 and 4.2.5 is the following extension of Theorem 3.5 in Dutta (1984) to topological effectivity functions.

**Corollary 4.2.6.** *Let $A$ be a normal space with rich structure $\mathcal{K}(A)$ and let $\Gamma = (N; \Sigma^1, \dots, \Sigma^n; g; A)$ be an acceptable game form on $V^N$. If $E^\Gamma$ is maximal, then $E^\Gamma$ is the effectivity function of a strong simple game.*

We conclude this section with the characterization of two-person acceptable game forms: under mild assumptions, the effectivity function of every two-person acceptable game form is dictatorial. A simple game $(N, \mathbf{W})$ is *dictatorial* if there is a $d \in N$ such that $\mathbf{W} = \{d\}^+$; $d$ is the *dictator*. An effectivity function is *dictatorial* if it is the effectivity function of a dictatorial simple game. We now extend Theorem 1 of Hurwicz and Schmeidler (1978) to topological effectivity functions.

**Theorem 4.2.7.** *Let $A$ be a normal space with rich structure $\mathcal{K}(A)$ and let $\Gamma = (N; \Sigma^1, \Sigma^2; g; A)$ be an acceptable two-person game form on $V^N$. If $E^\Gamma(i)$ is closed in the upper topology for $i = 1, 2$, then $E^\Gamma$ is dictatorial.*

To prove this theorem we need an auxiliary result. Let $E$ be an effectivity function and define $E^{**} = (E^*)^*$. Clearly, if $B \in E(S)$ and $B' \in E^*(N \setminus S)$, then $B' \cap B \neq \emptyset$ by definition of $E^*(N \setminus S)$; hence, $B \in (E^*)^*(S)$. So $E(S) \subseteq E^{**}(S)$. Moreover, we have the following lemma from Abdou and Keiding (1991, p. 46).

**Lemma 4.2.8.** *For every topological effectivity function $E$ and every $S \subseteq N$, $E^{**}(S)$ is the closure of $E(S)$ in the upper topology.*

*Proof of Theorem 4.2.7.* Since every two-person strong simple game is dictatorial, it is in view of Corollary 4.2.6 sufficient to prove that $E^\Gamma$ is maximal. Let $i \in \{1, 2\}$ then we have to prove that $\left(E^\Gamma\right)^*(i) = E^\Gamma(i)$. Clearly (as argued before, by superadditivity) $E^\Gamma(i) \subseteq \left(E^\Gamma\right)^*(i)$. Also, if $B \in \mathcal{K}(A)$ with $B \notin \left(E^\Gamma\right)^{**}(i)$, then there is $B' \in \left(E^\Gamma\right)^*(j)$ with $B \cap B' = \emptyset$, which in turn implies $B \notin \left(E^\Gamma\right)^*(i)$ by (3.7) of Theorem 3.4.3 – the intersection condition necessary for the existence of a Nash consistent representation. Hence, $\left(E^\Gamma\right)^*(i) \subseteq \left(E^\Gamma\right)^{**}(i)$.

By Lemma 4.2.8, $\left(E^\Gamma\right)^{**}(i) = \mathrm{cl}(E^\Gamma(i)) = E^\Gamma(i)$ (where 'cl' denotes the closure in the upper topology); the second equality follows since $E^\Gamma(i)$ is closed in the upper topology by assumption. Hence, $\left(E^\Gamma\right)^*(i) = E^\Gamma(i)$. □

## 4.3 Existence of acceptable representations

In the preceding section we have seen that two-person acceptable game forms are dictatorial. Therefore, in Sections 4.3–4.5 we assume that the number of players is at least three, $|N| = n \geq 3$. Moreover, we assume that $(A, \top)$ is a compact Hausdorff space where, as before, $\top$ denotes the topology on $A$. Recall (cf. Remark 4.2.2) that all our results hold for finite $A$ with structure $P_0(A)$ under the discrete topology.

Our main result is as follows.

**Theorem 4.3.1.** *Let $E : P(N) \to P(\mathcal{K}(A))$ be an effectivity function. Then $E$ has an acceptable representation on $V^N$ if and only if the following three conditions are satisfied.*

$$E \text{ is monotonic and superadditive.} \tag{4.6}$$

$$[S_i \in P_0(N),\ B_i \in E(S_i),\ i = 1, 2,\ S_1 \cap S_2 = \emptyset] \Rightarrow B_1 \cup B_2 = A\,. \tag{4.7}$$

*For every $R^N \in V^N$ there exists a Pareto optimal alternative $x \in A$ such that $L(x, R^N) \in E(N \setminus i)$ for all $i \in N$.* (4.8)

The necessity of (4.6) follows from Theorem 2.4.7 and the necessity of (4.7) follows from Theorem 4.2.3 (and the definition of a representation). Finally, the necessity of (4.8) follows from Proposition 3.2.1 and the definition of acceptability of a game form. The sufficiency part of Theorem 4.3.1 will be proved in Sections 4.4 and 4.5. Observe that actually (4.8) can be weakened by dropping the Pareto optimality requirement on $x \in A$: this follows from continuity of the preferences and compactness of closed subsets of $A$. Thus, by comparing Theorem 4.3.1 to Theorem 3.2.3 we see that the additional requirement of all Nash equilibrium outcomes being Pareto optimal is equivalent to the addition of condition (4.7).

By Corollary 3.4.11, we know that (4.8) is guaranteed by two other conditions, namely the familiar intersection conditions on polar sets and the condition that the sets assigned by the effectivity function to $(n-1)$-person coalitions are closed in the upper topology $\top_u$. Hence, we have the following corollary, which also follows by applying Corollary 3.6.3.

**Corollary 4.3.2.** *Let $E : P(N) \to P(\mathcal{K}(A))$ be an effectivity function. Then $E$ has an acceptable representation on $V^N$ if it satisfies* (4.6), (4.7), *and the following two assumptions:*

$$[B_i \in E^*(i) \text{ for all } i \in N] \Rightarrow \bigcap_{i=1}^{n} B_i \neq \emptyset\,. \tag{4.9}$$

$$E(N \setminus i) \text{ is closed in } (\mathcal{K}(A), \top_u) \text{ for every } i \in N\,. \tag{4.10}$$

Since (4.10) is automatically satisfied if $A$ is finite, and (4.9) is necessary for the existence of a Nash consistent representation (Theorem 3.4.3), we also have:

**Corollary 4.3.3.** *Let $2 \leq |A| < \infty$ and let $E : P(N) \to P(P_0(A))$ be an effectivity function. Then $E$ has an acceptable representation (on $W^N$) if and only if* (4.6), (4.7), *and* (4.9) *are satisfied.*

## 4.4 A game form with all Nash equilibrium outcomes Pareto optimal

In this section we make the first step towards constructing an acceptable game form in order to prove Theorem 4.3.1. We assume throughout that $A$ is a compact Hausdorff space, and that $N = \{1, \ldots, n\}$ with $n \geq 3$ is the set of players. Further, $E : P(N) \to P(\mathcal{K}(A))$ is an effectivity function satisfying (4.6) – monotonicity and superadditivity – and (4.7): for all $S_i \in P_0(N)$, $B_i \in E(S_i)$, $i = 1, 2$, and $S_1 \cap S_2 = \emptyset$, we have $B_1 \cup B_2 = A$. Our purpose is to construct a game form $\Gamma_1$ with the following two properties: (i) $\Gamma_1$ is a representation of $E$; and (ii) for every profile of continuous preferences $R^N \in V^N$, each Nash equilibrium of $(\Gamma_1, R^N)$ is Pareto optimal. This is not yet a proof of Theorem 4.3.1, since a game $(\Gamma_1, R^N)$ may fail to have a Nash equilibrium, i.e., $\Gamma_1$ is not Nash consistent. In Section 4.5 the game form $\Gamma_1$ will be extended to a game form $\Gamma_2$, under the additional assumption (4.8), which is an acceptable representation of $E$.

The game form $\Gamma_1$ will in fact be an 'extension' of the canonical game form $\Gamma_0$, constructed in the proof of Theorem 2.4.7 and used throughout Chapter 3. The structure is $\mathcal{T} = \mathcal{K}(A)$, which is rich since $A$ is Hausdorff. Let $\Gamma_0 = (N; \Sigma^1, \ldots, \Sigma^n; g_0; A)$, where, for each $i \in N$, $\Sigma^i = M^i \times \Phi \times N$. The set $M^i$ is defined by (2.7) and (2.8). The set of selections $\Phi$ is defined by $\Phi = \{\varphi : \mathcal{K}(A) \to A \mid \varphi(B) \in B \text{ for each } B \in \mathcal{K}(A)\}$. For the definition of the outcome function $g_0$ see the proof of Theorem 2.4.7.

Now the game form $\Gamma_1$ is the game form

$$\begin{aligned}\Gamma_1 &= (N; \Sigma^1 \times \{0,1\}, \ldots, \Sigma^n \times \{0,1\}; g_1; A) \\ &= (N; M^1 \times \Phi \times N \times \{0,1\}, \ldots, M^n \times \Phi \times N \times \{0,1\}; g_1; A) \\ &= (N; \Sigma_1^1, \ldots, \Sigma_1^n; g_1; A) ,\end{aligned}$$

where the outcome function $g_1$ is defined as follows. Let $\sigma = (\sigma^1, \ldots, \sigma^n)$ be an $n$-tuple of strategies for $\Gamma_1$, that is, $\sigma^i = (m^i, \varphi^i, t^i, q^i)$ for $i = 1, \ldots, n$, where $m^i \in M^i$, $\varphi^i \in \Phi$, $t^i \in N$, and $q^i \in \{0,1\}$. The equivalence relation $\sim_\sigma$ for each $S \in P_0(N)$ is the same as defined for $\Gamma_0$, see (2.9), and also the sequence of partitions is defined in the same way as in the proof of Theorem 2.4.7: after $r$ steps, we arrive at the finest possible partition $H_r(\sigma)$.

Let $H_r(\sigma) = \{S_1, \ldots, S_\ell\}$ and let $m_2^i(S_j) = B_j$, $j = 1, \ldots, \ell$, where $B_j \in E(S_j)$. So far we followed precisely the construction of $g_0$. Here, we deviate for the first time. Call a final coalition $S_j$, $1 \leq j \leq \ell$, *decided* if $q^i = 0$ for all $i \in S_j$; otherwise, $S_j$ is *undecided*. In defining $g_1$ we distinguish (without loss of generality) the following possibilities:

$S_1, \ldots, S_\ell$ are decided. (4.11)

In this case, we define $g_1(\sigma) = g_0(\sigma)$, thus: $g_1(\sigma) = \varphi^k(B_1 \cap \ldots \cap B_\ell)$, where $k = (t^1 + \cdots + t^n) \bmod n$ (observe that $B_1 \cap \ldots \cap B_\ell \neq \emptyset$ since $E$ is superadditive).

$S_1, \ldots, S_h$ are undecided, $1 \leq h \leq \ell$, and $S_{h+1}, \ldots, S_\ell$ are decided. (4.12)

For this case, in order to simplify notations, assume that $\bigcup_{j=1}^{h} S_j = \{1, \ldots, s\}$, where $1 \leq s \leq n$. Let $k = (t^1 + \cdots + t^s) \bmod s$. Then $g_1(\sigma) = \varphi^k(B_{h+1} \cap \ldots \cap B_\ell)$ if $h < \ell$ and $g_1(\sigma) = \varphi^k(A)$ if $h = \ell$. This completes the definition of $g_1$.

We now show that $\Gamma_1$ represents $E$ and that all Nash equilibrium outcomes of games associated with $\Gamma_1$ are Pareto optimal.

**Claim 4.4.1** *$\Gamma_1$ is a representation of $E$.*

*Proof.* Let $S \in P_0(N)$ and $B \in E(S)$. Let $\hat{m}^i(T) = (S, B)$ for all $T \supseteq S$ and $i \in S$. If $\sigma^i = (\hat{m}^i, \varphi^i, t^i, 0)$ for all $i \in S$, then for every $\tau^{N\setminus S} \in \Sigma_1^{N\setminus S}$, $S$ is a decided coalition with respect to $(\sigma^S, \tau^{N\setminus S})$. Hence, by (4.11) and (4.12), $g_1(\sigma^S, \tau^{N\setminus S}) \in B$ for all $\tau^{N\setminus S} \in \Sigma_1^{N\setminus S}$. Thus, $B \in E^{\Gamma_1}(S)$, and we have proved that $E^{\Gamma_1}(S) \supseteq E(S)$ for all $S \in P_0(N)$.

Now let $S \in P_0(N)$ and $C \in \mathcal{K}(A) \setminus E(S)$. Then $S \neq N$ since $E(N) = \mathcal{K}(A)$. In addition $B \setminus C \neq \emptyset$ for every $B \in E(S)$ since $E$ is monotonic. Let $\sigma^S \in \Sigma_1^S$ be fixed. We will choose strategies $\bar{\sigma}^i = (\bar{m}^i, \bar{\varphi}^i, \bar{t}^i, \bar{q}^i)$, $i \in N \setminus S$, such that $g_1(\sigma^S, \bar{\sigma}^{N\setminus S}) \notin C$, as follows. Let $\bar{m}^i(T) = (N \setminus S, A)$ for all $T \supseteq N \setminus S$ and $i \notin S$. Further let $\bar{q}^i = 1$ for all $i \in N \setminus S$. Then $N \setminus S$ is an undecided coalition with respect to $\sigma^* = (\sigma^S, \bar{\sigma}^{N\setminus S})$. If $S_1, \ldots, S_h$, $h \geq 0$,[3] are the decided coalitions in the final partition $H_r(\sigma^*)$, and $m_2^i(S_j) = B_j$, $j = 1, \ldots, h$, where $\sigma^i = (m^i, \varphi^i, t^i, q^i)$, $i \in S$, then $B := \bigcap_{j=1}^{h} B_j \in E(S)$. Let $k \in N \setminus S$. Player $k$ can choose $\widetilde{t}^k$ and $\widetilde{\varphi}^k$ such that $g_1(\sigma^*) = \widetilde{\varphi}^k(B) \notin C$. Thus, $C \notin E^{\Gamma_1}(S)$. □

**Claim 4.4.2** *Let $R^N \in V^N$. Then for every $\sigma \in NE(\Gamma_1, R^N)$, $g_1(\sigma)$ is Pareto optimal with respect to $R^N$.*

*Proof.* Let $\sigma = (\sigma^1, \ldots, \sigma^n)$ be a Nash equilibrium of $(\Gamma_1, R^N)$ and let the final partition of $N$ with respect to $\sigma$ be $\{S_1, \ldots, S_\ell\}$. Call $S_1, \ldots, S_\ell$ the final coalitions. We distinguish the following possible cases.

(i) *No final coalition is decided.* Let $x \in A$. Then, by (4.12), each player $i$ can make sure that $x$ is the outcome chosen by $g_1$, by choosing $t^i$ and $\varphi^i$ appropriately. Since $\sigma$ is a Nash equilibrium, we must therefore have $g_1(\sigma) R^i x$ for every $i \in N$. This holds for any $x \in A$, so $g_1(\sigma)$ is Pareto optimal.

(ii) *Exactly one final coalition is decided.* Let $S_j$ be the decided coalition. Then, by changing to $q^i = 1$ and choosing $t^i$ and $\varphi^i$ appropriately, each player $i \in S_j$ can make sure that $x \in A$ is chosen, for any $x \in A$. Since $\sigma$ is a Nash

[3] By $h = 0$ we indicate that there are no decided coalitions.

equilibrium, it follows that $g_1(\sigma)R^i x$ for all $i \in S_j$ and $x \in A$. Hence, $g_1(\sigma)$ is Pareto optimal.

(iii) *There exist at least two decided coalitions.* To simplify notations assume that $S_1, \dots, S_h$, $2 \le h \le \ell$, are the decided coalitions. Let $\sigma^i = (m^i, \varphi^i, t^i, q^i)$, $i \in N$, and let $B_j = m_2^i(S_j)$ for $i \in S_j$ and $j = 1, \dots, h$. Denote $C_j = \bigcap_{k \in \{1,\dots,h\} \setminus \{j\}} B_k$ for $j = 1, \dots, h$. Since player $i \in S_j$ can make coalition $S_j$ undecided and choose from $C_j$, and $\sigma$ is a Nash equilibrium, we must have $g_1(\sigma)R^i x$ for every $i \in S_j$ and $x \in C_j$, $j = 1, \dots, h$. We now assert that

$$\bigcup_{j=1}^{h} C_j = A\,. \tag{4.13}$$

We prove this assertion below, and first complete the proof of Claim 4.4.2. Let $x \in A$. By (4.13), $x \in C_j$ for some $j = 1, \dots, h$. Hence, $g_1(\sigma)R^i x$ for all $i \in S_j$. Thus, $g_1(\sigma)$ is Pareto optimal.

We are left to prove (4.13). Suppose, contrary to what we wish to prove, that there is $x \in A$ with $x \notin \bigcup_{j=1}^{h} C_j$. Since $x \notin C_1$ we can take $j \in \{2, \dots, h\}$ with $x \notin B_j$. Since $x \notin C_j$, we can take $k \in \{1, \dots, h\} \setminus \{j\}$ with $x \notin B_k$. Hence, $x \notin B_j \cup B_k$, contradicting (4.7). $\square$

## 4.5 Proof of Theorem 4.3.1

This section is completely devoted to the proof of Theorem 4.3.1, establishing the existence of acceptable representations under the conditions (4.6)–(4.8). In Section 4.4 we have defined the game form $\Gamma_1$ which represents $E$ and has only Pareto optimal Nash equilibrium outcomes, but does not have to be Nash consistent under the conditions (4.6)–(4.8). In this section we 'extend' $\Gamma_1$ to another game form $\Gamma_2$ that has all the desired properties.

We start by defining the social choice correspondence $H : V^N \to P_0(A)$ by

$$H(R^N) = \{a \in A \mid L(a, R^i) \in E(N \setminus i) \text{ for all } i \in N \text{ and } a \in PAR(R^N)\},$$

for all $R^N \in V^N$. By (4.8), $H$ is well defined, i.e., nonempty valued. Also, it is easy to see that $H$ is Maskin monotonic (see Remark 3.7.3). We denote

$$\text{graph}\,(H) = \{(R^N, a) \mid R^N \in V^N \text{ and } a \in H(R^N)\}\,,$$

and proceed to define $\Gamma_2 = (N; \Sigma_2^1, \dots, \Sigma_2^n; g_2; A)$, as follows. For each $i \in N$ let

$$\Sigma_2^i = \text{graph}\,(H) \times \{0,1\} \times \{0,1\} \times N \times E(i) \times \Phi \times \Sigma_1^i\,,$$

where $\Sigma_1^i$ is the strategy set of player $i$ in the game form $\Gamma_1$. It remains to define the outcome function $g_2$. Let

$$\eta^i = ((R^N)^i, a^i, q_1^i, q_2^i, t_0^i, B_0^i, \varphi_0^i, \sigma^i)$$

for each $i \in N$ describe an $n$-tuple $\eta = (\eta^1, \dots, \eta^n)$ of strategies. We distinguish the following possible cases.

$$\eta^i = (R^N, a, 0, 0, t_0^i, B_0^i, \varphi_0^i, \sigma^i) \text{ for all } i \in N. \tag{4.14}$$

In this case $g_2(\eta) = a$.

$$\begin{cases} \text{There exists } j \in N \text{ such that } \eta^i = (R^N, a, 0, 0, t_0^i, B_0^i, \varphi_0^i, \sigma^i) \\ \text{for all } i \in N \setminus j,\ ((R^N)^j, a^j, q_1^j, q_2^j) \neq (R^N, a, 0, 0), \text{ and } q_1^j = 0. \end{cases} \tag{4.15}$$

In this case, $g_2(\eta) = \varphi_0^j(L(a, R^j))$, where $R^j$ is the $j$-th component of $R^N$. (Observe that $g_2$ is well defined since $n \geq 3$.)

$$\begin{cases} \text{There exists } j \in N \text{ with } \eta^i = (R^N, a, 0, 0, t_0^i, B_0^i, \varphi_0^i, \sigma^i) \\ \text{for all } i \in N \setminus j, \text{ and } q_1^j = 1. \end{cases} \tag{4.16}$$

In this case, $g_2(\eta) = \varphi_0^j(L(a, R^j) \cap B_0^j)$. Note that the set $L(a, R^j) \cap B_0^j \neq \emptyset$ by superadditivity of $E$, since $L(a, R^j) \in E(N \setminus j)$ and $B_0^j \in E(j)$.

$$\begin{cases} \text{There exist } j, h \in N,\ j \neq h \text{ with } \eta^i = (R^N, a, 0, 0, t_0^i, B_0^i, \varphi_0^i, \sigma^i) \\ \text{for all } i \in N \setminus \{j, h\},\ \eta^h = (R^N, a, 0, 1, t_0^h, B_0^h, \varphi_0^h, \sigma^h), \\ ((R^N)^j, a^j, q_1^j, q_2^j) \neq (R^N, a, 0, 0), \text{ and } q_1^j = 0. \end{cases} \tag{4.17}$$

In this case, let $k = (t_0^1 + \ldots + t_0^n) \bmod n$. Then $g_2(\eta) = \varphi_0^k(A)$. (Observe that this case is always different from (4.15) since $n \geq 3$.)

$$\begin{cases} \text{There exist } j, h \in N,\ j \neq h \text{ with } \eta^i = (R^N, a, 0, 0, t_0^i, B_0^i, \varphi_0^i, \sigma^i) \\ \text{for all } i \in N \setminus \{j, h\},\ \eta^h = (R^N, a, 0, 1, t_0^h, B_0^h, \varphi_0^h, \sigma^h), \text{ and } q_1^j = 1. \end{cases} \tag{4.18}$$

In this case, $g_2(\eta) = \varphi_0^h(B_0^j)$.

*In all other cases*, let $g_2(\eta) = g_1(\sigma^1, \dots, \sigma^n)$, where $g_1$ is the outcome function of the game form $\Gamma_1$ of Section 4.4.

We will prove that $\Gamma_2$ is an acceptable representation of $E$. The proof is divided into several claims.

**Claim 4.5.1** *$\Gamma_2$ is Nash consistent.*

*Proof.* Let $R^N \in V^N$. Choose $a \in H(R^N)$ and define $\eta^i = (R^N, a, 0, 0, t_0^i, B_0^i, \varphi_0^i, \sigma^i)$ for all $i \in N$. Let $\eta = (\eta^1, \dots, \eta^n)$, then $\eta$ is a Nash equilibrium of $(\Gamma_2, R^N)$ by (4.14)–(4.16). □

**Claim 4.5.2** *For every $R^N \in V^N$, every Nash equilibrium outcome of $(\Gamma_2, R^N)$ is Pareto optimal.*

*Proof.* Let $R^N \in V^N$ and let $\eta = (\eta^1, \dots, \eta^n)$ be a Nash equilibrium of $(\Gamma_2, R^N)$. We distinguish the following cases, associated with (4.14)–(4.18) and the remaining cases for $g_2(\eta)$.

$$\eta^i = (\widehat{R}^N, \widehat{a}, 0, 0, t_0^i, B_0^i, \varphi_0^i, \sigma^i) \text{ for all } i \in N. \tag{4.14*}$$

Then $g_2(\eta) = \widehat{a}$ and $\widehat{a} \in H(\widehat{R}^N)$. By (4.15), $L(\widehat{a}, \widehat{R}^j) \subseteq L(\widehat{a}, R^j)$ for all $j \in N$. Thus, by Maskin monotonicity of $H$, $\widehat{a} \in H(R^N)$. Hence, $\widehat{a}$ is Pareto optimal with respect to $R^N$.

$$\begin{cases} \text{There exists } j \in N \text{ with } \eta^i = (\widehat{R}^N, \widehat{a}, 0, 0, t_0^i, B_0^i, \varphi_0^i, \sigma^i) \\ \text{for all } i \in N \setminus j,\ ((R^N)^j, a^j, q_1^j, q_2^j) \neq (\widehat{R}^N, \widehat{a}, 0, 0), \text{ and } q_1^j = 0. \end{cases} \quad (4.15^*)$$

Then, by (4.17), $A \subseteq L(g_2(\eta), R^i)$ for all $i \neq j$. Thus, $g_2(\eta)$ is Pareto optimal.

$$\begin{cases} \text{There exists } j \in N \text{ with } \eta^i = (\widehat{R}^N, \widehat{a}, 0, 0, t_0^i, B_0^i, \varphi_0^i, \sigma^i) \\ \text{for all } i \in N \setminus j, \text{ and } q_1^j = 1. \end{cases} \quad (4.16^*)$$

Then, by (4.15), $L(\widehat{a}, \widehat{R}^j) \subseteq L(g_2(\eta), R^j)$. By (4.18), $B_0^j \subseteq L(g_2(\eta), R^i)$ for all $i \neq j$. Also, by the definition of $H$, $L(\widehat{a}, \widehat{R}^j) \in E(N \setminus j)$. Thus, by (4.7), $L(\widehat{a}, \widehat{R}^j) \cup B_0^j = A$. So $g_2(\eta)$ is Pareto optimal.

$$\begin{cases} \text{There exist } j, h \in N,\ j \neq h \text{ with } \eta^i = (\widehat{R}^N, \widehat{a}, 0, 0, t_0^i, B_0^i, \varphi_0^i, \sigma^i) \\ \text{for all } i \in N \setminus \{j, h\},\ \eta^h = (\widehat{R}^N, \widehat{a}, 0, 1, t_0^h, B_0^h, \varphi_0^h, \sigma^h), \\ ((R^N)^j, a^j, q_1^j, q_2^j) \neq (\widehat{R}^N, \widehat{a}, 0, 0), \text{ and } q_1^j = 0. \end{cases} \quad (4.17^*)$$

Then, by (4.17), $A \subseteq L(g_2(\eta), R^i)$ for all $i \in N$. Hence, $g_2(\eta)$ is Pareto optimal.

$$\begin{cases} \text{There exist } j, h \in N,\ j \neq h \text{ with } \eta^i = (\widehat{R}^N, \widehat{a}, 0, 0, t_0^i, B_0^i, \varphi_0^i, \sigma^i) \\ \text{for all } i \in N \setminus \{j, h\},\ \eta^h = (\widehat{R}^N, \widehat{a}, 0, 1, t_0^h, B_0^h, \varphi_0^h, \sigma^h), \\ \text{and } q_1^j = 1. \end{cases} \quad (4.18^*)$$

Then, by (4.17), $A \subseteq L(g_2(\eta), R^i)$. Hence, $g_2(\eta)$ is Pareto optimal.

In all other cases, $g_2(\eta) = g_1(\sigma)$, where $\sigma$ is a Nash equilibrium of $(\Gamma_1, R^N)$. By the construction of $\Gamma_1$ in Section 4.4, $g_1(\sigma)$ is Pareto optimal. □

Our final claim completes the proof of Theorem 4.3.1.

**Claim 4.5.3** *$\Gamma_2$ is a representation of $E$.*

*Proof.* First consider $S \subseteq N$ with $|S| \geq 2$. Since $S$ can enforce the play of the game to be in $\Gamma_1$, we have $E^{\Gamma_1}(S) \subseteq E^{\Gamma_2}(S)$. By going over cases (4.14)–(4.18), it follows that $S$ cannot enforce the outcome to be in any set that is not already in $E(S) = E^{\Gamma_1}(S)$. Hence, $E^{\Gamma_2}(S) \subseteq E^{\Gamma_1}(S)$. So $E^{\Gamma_2}(S) = E^{\Gamma_1}(S) = E(S)$.

We still have to prove that $E^{\Gamma_2}(i) = E(i)$ for every $i \in N$. Let $i \in N$ and $B \in E(i)$. Consider the following strategy of player $i$: $\eta^i = ((R^N)^i, a^i, 1, q_2^i, t_0^i, B, \varphi_0^i, \sigma^i)$, where $\sigma^i = (m^i, \varphi^i, t^i, 0)$ and $m^i(S) = (\{i\}, B)$ for all $S \in N^i$. It is easy to verify that $g_2(\eta^i, \eta^{N\setminus i}) \in B$ for all $\eta^{N\setminus i} \in \Sigma_2^{N\setminus i}$. Thus, $B \in E^{\Gamma_2}(i)$ and $E(i) \subseteq E^{\Gamma_2}(i)$. Now suppose that $B' \in E^{\Gamma_2}(i)$. As $|N \setminus i| \geq 2$, $N \setminus i$ can enforce the play of $\Gamma_1$. So $B' \in E^{\Gamma_1}(i)$. Thus, $E^{\Gamma_2}(i) \subseteq E^{\Gamma_1}(i) = E(i)$. □

## 4.6 Notes and comments

This chapter is based on Peleg (2004). It is interesting to note that Theorem 4.3.1 applies to the following family of effectivity functions. (A *vetoer* in a simple game – see Example 3.5.1 – is a player who belongs to each winning coalition.)

**Proposition 4.6.1.** *Let $G$ be a proper simple game with at most one vetoer. Then the associated effectivity function $E(G) : P(N) \to P(\mathcal{K}(A))$, where $A$ is a compact Hausdorff space, has an acceptable representation on $V^N$.*

*Proof.* Since $G$ is proper and (by definition) monotonic, $E(G)$ satisfies (4.6) and (4.7). In addition, if $i \in N$ is not a vetoer then $N \setminus i$ is winning, so $E(N \setminus i) = \mathcal{K}(A)$ and therefore $E^*(i) = \{A\}$. Since there is at most one vetoer, (4.9) is satisfied. Also, for each $B \in \mathcal{K}(A) \setminus \{A\}$ we have $B \subseteq A \setminus \{x\}$ for some $x \in A \setminus B$, and $A \setminus \{x\}$ is open since $\{x\} \in \mathcal{K}(A)$. Thus, $\mathcal{K}(A) \setminus \{A\}$ is open and hence $\{A\}$ is closed in the upper topology. Furthermore, $\mathcal{K}(A)$ is closed in the upper topology. Hence, (4.10) is satisfied. The proof is complete by applying Corollary 4.3.2. □

In particular, as mentioned before, simple majority committees (i.e., games) have acceptable representations.

*Remark 4.6.2.* As was already noted, but nevertheless is remarkable, (4.8) can be replaced by the weaker condition (3.3), namely: for every $R^N \in V^N$ there exists $x \in A$ such that $L(x, R^i) \in E(N \setminus i)$ for all $i \in N$. This follows from the fact that preferences are continuous and every closed set is compact. Alternatively, one can apply Theorems 3.4.10 and 3.2.3.

# Chapter 5
# Strongly consistent representations

## 5.1 Motivation and summary

In the preceding chapters we have studied representations of constitutions (effectivity functions) under a minimal stability condition, namely Nash consistency: for every admissible profile of preferences there should be at least one Nash equilibrium in the representing game. In Chapter 4 we have studied acceptable representations, meaning that all Nash equilibria are Pareto optimal. Another way to look at this is that not only single players do not wish to deviate, but also the grand coalition of all players has no incentive to jointly deviate to a different strategy profile. More generally, one may argue that in many interesting conflict situations preplay communication, direct or indirect, is possible – and sometimes unavoidable. This leads naturally to coordination of strategies by coalitions of players and may upset a given Nash equilibrium. To avoid this and maintain stability, we need to consider coalitional equilibrium concepts where coalitions have no profitable deviations. In this chapter we consider the strongest equilibrium concept, namely *strong equilibrium*. In a strong equilibrium no coalition $S$ of players has a deviating $S$-tuple of strategies that is (strictly) profitable for each of its members – see Definition 5.2.1. A game form $\Gamma$ is strongly consistent if for every profile of preferences $R^N$ the resulting game $(\Gamma, R^N)$ has a strong equilibrium. The main goal in the chapter is to characterize effectivity functions that can be represented by strongly consistent game forms. An effectivity function is *stable* if its *core* is nonempty for every profile of preferences – see Definition 5.2.5. Our main result is the following: an effectivity function has a strongly consistent representation if and only if it is maximal and stable. Although stability is equivalent to *acyclicity* (Keiding, 1985), a condition which is formulated directly on effectivity functions, we have also a simpler characterization, namely: an effectivity function has a strongly consistent representation if and only if it is maximal and *convex*. Convexity of an effectivity function can usually easily be checked.

B. Peleg and H. Peters, *Strategic Social Choice*, Studies in Choice and Welfare, 
DOI 10.1007/978-3-642-13875-1_5, 

In Sections 5.2 and 5.3 we study the mentioned necessary and sufficient conditions for the existence of strongly consistent representations in case the set of alternatives is finite. Section 5.4 extends these results to topological effectivity functions.

## 5.2 Necessary conditions for strongly consistent representations

Throughout this section we assume that the set of alternatives $A$ is finite, with $|A| \geq 2$.

When preplay communication is possible, coalitions of players may coordinate their strategies and possibly upset a Nash equilibrium. Consider for instance the 'kingmaker game form' (Examples 2.4.2, 2.4.4) of which the associated effectivity function $E$ assigns $P_0(A)$ to every two-person coalition and $\{A\}$ to every one-person coalition, where $A = \{a, b, c\}$. Hence, $E = E(G)$, where $G$ is the three-person simple majority game, and the well known 'voting paradox' applies. Specifically, let $R^1 = (a, b, c)$, $R^2 = (c, a, b)$, and $R^3 = (b, c, a)$, then for any $x \in A$ there exists $y \in A$ and $i, j \in N$, $i \neq j$, such that $yP^ix$ and $yP^jx$. Hence, in any representing game form coalition $\{i, j\}$ can upset a Nash equilibrium with outcome $x$. For an effectivity function like the one associated with the 'kingmaker game form' – individuals have no rights and $(n-1)$-coalitions are all powerful – existence of Nash consistent representations is obvious[1], but strategic behavior of coalitions may upset Nash equilibria.

Examples like this one lead to the following definition.

**Definition 5.2.1.** Let $\Gamma = (N; \Sigma^1, \ldots, \Sigma^n; g; A)$ be a game form and let $R^N \in W^N$. Strategy profile $\sigma = (\sigma^1, \ldots, \sigma^n) \in \Sigma$ is a *strong (Nash) equilibrium* (SNE) of $(\Gamma, R^N)$ if for every $S \in P_0(N)$ and every $\mu^S \in \Sigma^S$, there exists $i \in S$ such that $g(\sigma)R^ig(\mu^S, \sigma^{N\setminus S})$. Game form $\Gamma$ is *strongly consistent* if for every $R^N \in W^N$ the game $(\Gamma, R^N)$ has at least one strong equilibrium.

An example of a strongly consistent game form is the following.

*Example 5.2.2.* Let $N = \{1, \ldots, n\}$, $n \geq 2$, $A = \{a_1, \ldots, a_m\}$, $m \geq 3$, and $\Sigma^1 = \Sigma^2 = \ldots = \Sigma^n = W$. It remains to define $g$. (Note that $g : W^N \to A$ is actually a *social choice function.*) Let $R^N \in W^N$. As before, denote by $PAR(R^N)$ the set of Pareto optimal alternatives with respect to $R^N$. If $a_1 \in PAR(R^N)$ we define $g(R^N) = a_1$. Otherwise, we define $g(R^N) = a_k$, where $a_k$ is the first alternative (in the order $a_1, a_2, \ldots$) in $PAR(R^N)$ that Pareto dominates $a_1$. We claim that $\Gamma = (N; \Sigma^1, \ldots, \Sigma^n; g; A)$ is strongly consistent. To prove this we distinguish the following possibilities. Let $R^N \in W^N$.

[1] By Proposition 3.2.2 every $x \in A$ is a Nash equilibrium outcome of $(\Gamma_0, R^N)$ for every $R^N \in W^N$.

(i) $a_1 \in PAR(R^N)$. For every $i \in N$ let $Q^i$ be the preference $Q^i = (a_1, \ldots, a_n)$, i.e., $a_1 Q^i \ldots Q^i a_n$. Then $Q^N$ is a strong Nash equilibrium of $(\Gamma, R^N)$ since no coalition $S$, $S \neq \emptyset, N$ can change the outcome $g(Q^N) = a_1$, and $N$ cannot change the outcome profitably, as $a_1 \in PAR(R^N)$.

(ii) $a_1 \notin PAR(R^N)$. Let $a_k \in PAR(R^N)$ be the first alternative that Pareto dominates $a_1$. Without loss of generality $k = 2$. Define $Q^i = (a_2, a_1, a_3, \ldots, a_n)$ for each $i \in N$. Then $Q^N$ is a strong Nash equilibrium of $(\Gamma, R^N)$. Indeed, if $S \in P_0(N)$, $S \neq N$, and $\widetilde{Q}^S \in W^S$, then $g(\widetilde{Q}^S, Q^{N \setminus S}) \in \{a_1, a_2\}$. Since $a_2$ Pareto dominates $a_1$ according to $R^N$, this implies that $S$ cannot profitably deviate. Since $a_2$ is Pareto optimal according to $R^N$, also $N$ cannot profitably deviate. (This example is due to Dutta and Pattanaik (1978).)

For easy reference, we state in the following remark a simple consequence of the definition of the effectivity function associated with a game form (see Definition 2.4.3).

*Remark 5.2.3.* Let $\Gamma = (N; \Sigma^1, \ldots, \Sigma^n; g; A)$ be a game form, let $S \in P_0(N)$, $S \neq N$, and $B \notin E^\Gamma(S)$. Then for every $\sigma^S \in \Sigma^S$ there exists $\mu^{N \setminus S} \in \Sigma^{N \setminus S}$ such that $g(\sigma^S, \mu^{N \setminus S}) \in A \setminus B$.

Our first result gives a necessary condition for representation of effectivity functions by strongly consistent game forms, namely maximality of the effectivity function. Recall (see Remark 3.7.2) that an effectivity function $E$ is maximal if it is superadditive and equal to its polar, i.e., $E = E^*$. See also the discussion in Section 4.2.

**Proposition 5.2.4.** *Let $E : P(N) \to P(P_0(A))$ be an effectivity function that has a strongly consistent representation. Then $E$ is maximal.*

*Proof.* Let $\Gamma = (N; \Sigma^1, \ldots, \Sigma^n; g; A)$ be a strongly consistent representation of $E$. Then $E = E^\Gamma$ is superadditive. Suppose, on the contrary, that $E$ is not maximal. Then there is a coalition $S \in P_0(N)$, $S \neq N$, and a set of alternatives $B \in E^*(S) \setminus E(S)$. Since $B \in E^*(S)$, we must have $A \setminus B \notin E(N \setminus S)$, as $B \cap (A \setminus B) = \emptyset$. Consider a profile $R^N \in W^N$ with $BP^S(A \setminus B)$ and $(A \setminus B)P^{N \setminus S}B$, i.e, each player in $S$ strictly prefers each element of $B$ to each element of $A \setminus B$, and each player in $N \setminus S$ strictly prefers each element of $A \setminus B$ to each element of $B$.[2] We claim that $(\Gamma, R^N)$ has no strong Nash equilibrium. Indeed, let $\sigma \in \Sigma$. If $g(\sigma) = x \in B$, then by Remark 5.2.3 coalition $N \setminus S$ has a strategy combination $\mu^{N \setminus S} \in \Sigma^{N \setminus S}$ such that $g(\sigma^S, \mu^{N \setminus S}) = y \in A \setminus B$. Since $yP^i x$ for all $i \in N \setminus S$, coalition $N \setminus S$ can profitably deviate from $\sigma$. Similarly, if $g(\sigma) \in A \setminus B$ then $S$ can profitably deviate from $\sigma$. Hence, $(\Gamma, R^N)$ has no strong Nash equilibrium, which is the desired contradiction. □

[2] So $P^i$ denotes the asymmetric part of $R^i$ for each $i \in N$.

In order to formulate our second necessary condition we need the (general) definition of the core of an effectivity function. (This concept was introduced earlier – Definition 3.3.3 – for the residual of an effectivity function.)

**Definition 5.2.5.** Let $E : P(N) \to P(P_0(A))$ be an effectivity function and let $R^N \in W^N$. We say that $x \in A$ is *dominated at* $R^N$ if there exist $S \in P_0(N)$ and $B \in E(S)$ such that $x \notin B$ and $BP^S x$. The *core* of $E$ with respect to $R^N$, denoted by $C(E, R^N)$, is the set of undominated alternatives with respect to $R^N$. The effectivity function is *stable* if $C(E, R^N) \neq \emptyset$ for every $R^N \in W^N$.

The second necessary condition for the existence of a strongly consistent representation is stability of the effectivity function.

**Proposition 5.2.6.** *Let* $E : P(N) \to P(P_0(A))$ *be an effectivity function that has a strongly consistent representation* $\Gamma = (N; \Sigma^1, \ldots, \Sigma^n; g; A)$. *Then* $g(\sigma) \in C(E, R^N)$ *for every strong equilibrium* $\sigma$ *of* $(\Gamma, R^N)$, *for every* $R^N \in W^N$. *In particular,* $E$ *is stable.*

*Proof.* Let $R^N \in W^N$ and let $\sigma$ be a strong equilibrium of the game $(\Gamma, R^N)$. We claim that $x = g(\sigma) \in C(E, R^N)$. Indeed, suppose to the contrary that $x$ is dominated at $R^N$. Then there are $S \in P_0(N)$ and $B \in E(S)$ with $x \notin B$ and $BP^S x$. Since $B \in E(S)$, $S$ has a strategy combination $\mu^S \in \Sigma^S$ such that $g(\mu^S, \tau^{N \setminus S}) \in B$ for all $\tau^{N \setminus S} \in \Sigma^{N \setminus S}$. In particular, $y = g(\mu^S, \sigma^{N \setminus S}) \in B$. Thus, $yP^i x$ for all $i \in S$, so that $S$ has a profitable deviation from $\sigma$, a contradiction. □

In the next section we prove the converse result: a maximal and stable effectivity function has a strongly consistent representation.

## 5.3 Existence of strongly consistent representations

We assume the same framework as in the preceding section. So $A$ is a finite set of at least two alternatives and $E : P(N) \to P(P_0(A))$ is an effectivity function. For completeness and easy reference we state the following observation.

*Remark 5.3.1.* If $E$ is maximal then $E$ is monotonic. Indeed, if $E$ is maximal then $E = E^*$, so that $E$ is monotonic with respect to alternatives since $E^*$ is. Also, since $E$ is superadditive (by definition of maximality), $E$ is monotonic with respect to coalitions (see the paragraph following Definition 2.3.5 for an exact argument).

We now show that maximality and stability of the effectivity function are sufficient for the existence of a strongly consistent representation.

**Theorem 5.3.2.** *Let the effectivity function $E : P(N) \to P(P_0(A))$ be stable and maximal. Then $E$ has a strongly consistent representation.*

*Proof.* By Remark 5.3.1 and Theorem 2.4.7, the canonical game form $\Gamma_0 = (N; \Sigma^1, \ldots, \Sigma^n; g_0; A)$ constructed in the proof of that theorem, is a representation of $E$. We show that $\Gamma_0$ is strongly consistent.

Let $R^N \in W^N$ and let $a \in C(E, R^N)$. For each $T \in P_0(N)$ we denote by

$$Pr(T, a, R^N) = \{b \in A \mid bP^i a \text{ for all } i \in T\} \tag{5.1}$$

the set of all alternatives strictly preferred to $a$ by all players in $T$ according to $R^N$. Since $a \in C(E, R^N)$, we have $Pr(T, a, R^N) \notin E(T)$. Hence, because of maximality of $E$,

$$A \setminus Pr(T, a, R^N) \in E(N \setminus T) . \tag{5.2}$$

Using (5.2) we define an $n$-tuple of strategies $\sigma = (\sigma^1, \ldots, \sigma^n) = ((m^1, \varphi^1, t^1), \ldots, (m^n, \varphi^n, t^n))$ in $\Gamma_0$ by

$$m^i(T) = \begin{cases} (T, A \setminus Pr(N \setminus T, a, R^N)) & \text{if } T \in P_0(N),\ T \neq N,\ i \in T \\ (N, \{a\}) & \text{if } T = N \end{cases} \tag{5.3}$$

and $\varphi^i, t^i$, $i \in N$ arbitrary but fixed. Note that the strategies $\sigma^i$ are well defined, in particular, each $m^i$ satisfies the monotonicity condition (2.8). Indeed, if $T_1 \subseteq T_2$ then $N \setminus T_1 \supseteq N \setminus T_2$, hence $Pr(N \setminus T_1, a, R^N) \subseteq Pr(N \setminus T_2, a, R^N)$, so that $A \setminus Pr(N \setminus T_1, a, R^N) \supseteq A \setminus Pr(N \setminus T_2, a, R^N)$.

We claim that $\sigma$ is a strong Nash equilibrium of $(\Gamma_0, R^N)$. Note that $g(\sigma) = a$. Let $S \in P_0(N)$ and $\mu^S \in \Sigma^S$. It is sufficient to prove that

$$g_0(\sigma^{N \setminus S}, \mu^S) \notin Pr(S, a, R^N) . \tag{5.4}$$

For $S = N$, (5.4) follows from the Pareto optimality of $a$, which in turn follows since $a \in C(E, R^N)$. Now let $S \neq N$. Let $\{T_1, \ldots, T_\ell\} = H_r(\sigma^{N \setminus S}, \mu^S)$ be the final partition associated with $(\sigma^{N \setminus S}, \mu^S)$. By (5.3) and the definition of $\sigma^i$, the members of $N \setminus S$ are not separated in $H_r(\sigma^{N \setminus S}, \mu^S)$, that is, there exists $1 \leq j \leq \ell$ such that $N \setminus S \subseteq T_j$ and $m^i_2(T_j) = A \setminus Pr(N \setminus T_j, a, R^N)$ for all $i \in T_j$. Thus, $g_0(\sigma^{N \setminus S}, \mu^S) \in A \setminus Pr(N \setminus T_j, a, R^N)$. Since $S = N \setminus (N \setminus S) \supseteq N \setminus T_j$, we have $Pr(S, a, R^N) \subseteq Pr(N \setminus T_j, a, R^N)$. So $A \setminus Pr(S, a, R^N) \supseteq A \setminus Pr(N \setminus T_j, a, R^N)$. Therefore, $g(\sigma^{N \setminus S}, \mu^S) \in A \setminus Pr(S, a, R^N)$, which implies (5.4). □

Stability of an effectivity function can be expressed directly, in terms of the effectivity function itself. Keiding (1985) characterized stability by a finite and explicit – but rather complex – condition of acyclicity. Fortunately the combination of maximality and stability can be checked in a much simpler way. We first introduce the following definition, which strengthens the condition of superadditivity.

**Definition 5.3.3.** An effectivity function $E : P(N) \to P(P_0(A))$ is *convex* if

$$[B_i \in E(S_i),\ i = 1, 2] \Rightarrow B_1 \cap B_2 \in E(S_1 \cup S_2) \text{ or } B_1 \cup B_2 \in E(S_1 \cap S_2) . \tag{5.5}$$

From Peleg (1984) we quote the following results: (i) A maximal and stable effectivity function is convex (Theorem 6.A.9). (ii) A convex effectivity function is stable (Theorem 6.A.7). With these results we obtain the following corollary.

**Corollary 5.3.4.** *An effectivity function* $E : P(N) \to P(P_0(A))$ *has a strongly consistent representation if and only if it is convex and maximal.*

Thus, in order to verify existence of strongly consistent representations only the (relatively) simple conditions of maximality and convexity have to be checked.

We conclude with the following example.

*Example 5.3.5.* Let $|N|, |A| \geq 2$ and let $p$ and $q$ be strictly positive probability vectors on $A$ and $N$, respectively. Define the (additive) effectivity function $E$ by

$$B \in E(S) \Leftrightarrow q(S) > 1 - p(B) .$$

Then $E$ is a convex effectivity function, as is straightforward to verify. If, in addition, $q(S) \neq p(B)$ for all $S \in P_0(N)$, $S \neq N$, and $B \in P_0(A)$, $B \neq A$, then $E$ is also maximal. This is again easy to verify – see also Lemma 6.2.51 in Peleg (1984).

## 5.4 Strongly consistent representations of topological effectivity functions

In this section $A$ is a topological space and $E : P(N) \to P(\mathcal{K}(A))$ a topological effectivity function. Before stating and proving our main result we introduce and recall some definitions.

A set $B \subseteq A$ is a $G_\delta$ if it is the intersection of a countable family of open sets in $A$. We shall use the following auxiliary result.

**Lemma 5.4.1.** *Let* $A$ *be a normal topological space and let* $B \subseteq A$ *be a closed* $G_\delta$. *Then there exists a continuous function* $f : A \to [0, 1]$ *such that* $B = f^{-1}(0)$.

See Kelley (1955, p. 134). The lemma applies in particular if $A$ is a compact Hausdorff space.

The definitions of stability and strongly consistent representation of a topological effectivity function are straightforward adaptations of those in the finite case. In our characterization theorem we will use the Vietoris topology on $\mathcal{K}(A)$, which is finer than the upper topology: see Remark 3.7.5 for the definition.

**Theorem 5.4.2.** *Let* $A$ *be a compact Hausdorff space such that each closed set is a* $G_\delta$. *Then:*

(*i*) *If* $E : P(N) \to P(\mathcal{K}(A))$ *has a strongly consistent representation, then* $E$ *is stable and maximal.*
(*ii*) *If* $E : P(N) \to P(\mathcal{K}(A))$ *is stable and maximal and* $E(S)$ *is closed in the Vietoris topology for every* $S \in P_0(N)$, *then* $E$ *has a strongly consistent representation.*

*Proof.* (i) Let $\Gamma = (N; \Sigma^1, \ldots, \Sigma^n; g; A)$ be a strongly consistent representation of $E$. For the stability of $E = E^\Gamma$ the proof of Proposition 5.2.6 can be copied. To prove maximality assume, per absurdum, that there exists $S \in P_0(N)$, $S \neq \emptyset$, and $B \in E^*(S) \setminus E(S)$. By Lemma 5.4.1 there exist continuous functions $u_1, u_2 : A \to [0,1]$ with $u_1^{-1}(1) = B$ and $u_2^{-1}(0) = B$. Consider the profile $u^N \in V^N$ with $u^i = u_1$ for each $i \in S$ and $u^i = u_2$ for each $i \in N \setminus S$. We claim that $(\Gamma, u^N)$ has no strong Nash equilibrium. Indeed, let $\sigma = (\sigma^1, \ldots, \sigma^n) \in \Sigma$ and let $x = g(\sigma)$. If $x \in B$ then $u^i(x) = 0$ for all $i \in N \setminus S$. As $B \notin E(S)$, $N \setminus S$ has a strategy $\mu^{N \setminus S} \in \Sigma^{N \setminus S}$ such that $g(\sigma^S, \mu^{N \setminus S}) = y \in A \setminus B$. Thus, $u^i(y) > 0 = u^i(x)$ for all $i \in N \setminus S$, and therefore $\mu^{N \setminus S}$ is an improvement of $N \setminus S$ on $\sigma$. Hence, in this case $\sigma$ is not a strong Nash equilibrium.

If $x \in A \setminus B$ then $u_1(x) < 1$. Consider the set $D = \{z \in A \mid u_1(z) \leq \frac{1+u_1(x)}{2}\}$. Then by continuity of $u_1$ the set $D$ is a closed subset of $A \setminus B$. Since $B \in E^*(S)$, $D \notin E(N \setminus S)$. Hence there exists $\mu^S \in \Sigma^S$ such that $g(\mu^S, \sigma^{N \setminus S}) = y \notin D$. Thus, $u^i(y) > \frac{1+u_1(x)}{2} > u_1(x)$ for all $i \in S$, so that $\mu^S$ is an improvement of $S$ on $\sigma$. Hence, also in this case $\sigma$ is not a strong Nash equilibrium.

(ii) The proof of this part is a modification of the proof of Theorem 5.3.2, and uses again the canonical representation from the proof of Theorem 2.4.7. It is sufficient to prove the following claim, the rest of the proof is a copy of the proof of Theorem 5.3.2.

*Claim.* Let $R^N \in V^N$ be a profile of (continuous) preferences, and let $a \in C(E, R^N)$. For $T \in P_0(N)$ denote

$$Pr(T, a, R^N) = \{b \in A \mid bP^i a \text{ for all } i \in T\}\ .$$

Then $A \setminus Pr(T, a, R^N) \in E(N \setminus T)$.

*Proof of the Claim.* We may assume $Pr(T, a, R^N) \neq \emptyset$ since otherwise we are done. Since $A \setminus Pr(T, a, R^N)$ is closed, by Lemma 5.4.1 there exists a continuous function $u : A \to [0,1]$ such that for all $x \in A$ we have $u(x) = 0 \Leftrightarrow x \in A \setminus Pr(T, a, R^N)$. Consider the sets

$$D_k = \{x \in A \mid u(x) \geq \frac{1}{k}\}, \quad k = 2, 3, \ldots$$

Clearly, $D_k \subseteq Pr(T, a, R^N)$ for all $k = 2, 3, \ldots$ As $u(a) = 0$ and $a \in C(E, R^N)$, $D_k \notin E(T)$. Since $E(T) = E^*(T)$ by maximality of $E$, there exists $\widetilde{D}_k \in E(N \setminus T)$ such that $\widetilde{D}_k \cap D_k = \emptyset$, for all $k$. By Klein and Thompson (1984, Theorem 2.3.5(iii)), $E(N \setminus T)$ is compact in the Vietoris topology. Thus there is a subnet of $\{\widetilde{D}_k \mid k = 2, 3, \ldots\}$ converging to some

$\widetilde{D} \in E(N \setminus T)$. Clearly, $\widetilde{D} \subseteq A \setminus Pr(T, a, R^N)$. By monotonicity of $E$, $A \setminus Pr(T, a, R^N) \in E(N \setminus T)$. □

Like in the finite case, if $E$ is convex, then $E$ is stable since $A$ is compact (see Abdou and Keiding, 1991). Thus, convexity may replace stability in (ii) of Theorem 5.4.2.

Observe, further, that the theorem applies in particular to compact metric spaces.

## 5.5 Notes and comments

Strongly consistent representations of effectivity functions were first mentioned in Remark 4.4 of Peleg (1998). However, strongly consistent representations of simple games were considered much earlier (see Peleg, 1978b, and Holzman, 1986a,b). In the earlier representations the representing game forms were actually social choice functions, i.e., functions from the set of profiles of preferences to the set of alternatives. We will see such a representation in Part II of this book.

Stability of effectivity functions was first considered in Moulin and Peleg (1982), where the non-emptiness of the core of an additive effectivity function had been proved – see Peleg (1984, p. 126) for the definition of an additive effectivity function. That result was followed by Peleg's (1984) proof of the stability of convex effectivity functions. The complete solution to the stability problem with finitely many alternatives was given in Keiding (1985). This was followed by some works of Abdou and Keiding on stability of cores of effectivity functions in topological spaces – see Abdou and Keiding (1991) for a presentation of the results.

# Chapter 6
# Nash consistent representation through lottery models

## 6.1 Motivation and summary

In Chapter 3 we have seen that – under the usual assumptions of monotonicity and superadditivity, and for a finite set of alternatives (social states) – effectivity functions (constitutions) can be represented by Nash consistent game forms if and only if the intersection condition on individual polar sets (3.6) is satisfied. This condition is quite restrictive, for instance, it is not satisfied by the effectivity function derived from the familiar $2 \times 2$ bimatrix game form (cf. Example 3.3.11).

A well-known way to avoid this condition is offered by game theory. If we represent preferences by von Neumann-Morgenstern utilities and allow mixed strategies in the representing game form, e.g., the canonical game form $\Gamma_0$ constructed in the proof of Theorem 2.4.7, then there always exists a Nash equilibrium in mixed strategies (Nash, 1951). Mixed strategies, however, can be hard to interpret – this is a longstanding discussion in game theory that we do not want to enter into here. Moreover, also the representation issue is under discussion, since outcomes can be probability distributions over the alternatives resulting from the use of mixed strategies, rather than only the original pure alternatives. In other words, admitting mixed strategies implies admitting lotteries over outcomes.

In this chapter we follow a different route in order to avoid the intersection condition (3.6). Instead of allowing mixed strategies we allow for some objective uncertainty concerning the outcomes of the game form. That is, we allow (some) lotteries over outcomes but do not need to allow mixed strategies. Specifically, we add the (finite) set of equal chance lotteries over the alternatives to the set of outcomes. Of course, also in this approach we have to look closely at the representation issue: the effectivity function associated with such a game form assigns sets of lotteries to coalitions. We will handle this question by considering so called lottery models. If $E$ is the original effectivity function and $\widetilde{E}$ is the effectivity function associated with the game

B. Peleg and H. Peters, *Strategic Social Choice*, Studies in Choice and Welfare,
DOI 10.1007/978-3-642-13875-1_6, © Springer-Verlag Berlin Heidelberg 2010

form $\widetilde{\Gamma}$ augmented by lotteries, then we will say that $\widetilde{E}$ *is a lottery model for* $E$ if the following holds: for each coalition $S$ and each $B \in E(S)$ there is a $\widetilde{B} \in \widetilde{E}(S)$ such that the total support (i.e., union of the supports) of the lotteries in $\widetilde{B}$ is equal to $B$; and, conversely, if $\widetilde{B} \in \widetilde{E}(S)$, then $B \in E(S)$ where $B$ is the total support of the lotteries in $\widetilde{B}$. This seems a natural way of extending the idea of representation to lotteries. If a coalition $S$ is effective for a set $B$ then this means that $S$ is entitled to or can enforce the final outcome to be in $B$; the same holds if $S$ is effective for a set of lotteries with total support equal to $B$.

In the augmented game form, we assume that players evaluate such lotteries by utility functions satisfying the minimal requirement of stochastic dominance: this means that utility increases by shifting probability to better (pure) alternatives. Thus, if lottery $\ell'$ can be obtained from lottery $\ell$ by shifting probability to more preferred alternatives, then $\ell'$ is preferred over $\ell$. Expected utility, for instance, is a special case of this.

With these assumptions, we are able to prove that for any effectivity function (satisfying the usual necessary conditions of monotonicity and superaddivity, and for a finite set of alternatives) there exists a lottery model which has a Nash consistent representation, without imposing further conditions on the effectivity function. The representing game form is finite, and no mixed strategies are used. The players play pure strategies, but the outcome may be uncertain.

As a simple but illustrative example, consider the unanimity effectivity function, where the grand coalition is effective for every single alternative and all other coalitions are completely powerless, i.e., only effective for the set of all alternatives. Since in any representing game form of this effectivity function any individual can bring about any alternative, given the strategy profile of the coalition of all other players, it follows that for any profile of preferences in which at least two players have different top elements, a Nash equilibrium cannot exist. By extending the effectivity function with equal chance lotteries, for instance such that every coalition other than the grand coalition is effective for every set of lotteries containing the lottery that assigns equal probability to each alternative, we obtain an effectivity function that preserves power and does have a Nash consistent representing game form – see Example 6.3.4 below.

For a constitution modeled as a monotonic and superadditive effectivity function, the relevance of our main result (Theorem 6.3.2) is that such a constitution can always be 'decentralized' by a set of rules (a game form) that preserves the original rights and that is stable in the sense that for any preferences a Nash equilibrium exists, as long as we are willing to accept some uncertainty in the form of equal chance lotteries as outcomes of the game, evaluated by utility functions respecting stochastic dominance.

For a given finite game form our result implies that we can always find an alternative finite game form, preserving effectivity in the indicated sense, that has a pure Nash equilibrium for any profile of preferences, again evaluating lotteries by utility functions respecting stochastic dominance (e.g., expected utility). This also entails a solution to the Gibbard (1974) paradox – see Example 3.3.11 and Example 6.3.3 below.

In Section 6.2 we extend Theorem 3.3.10 to accommodate for cardinal utilities, which are used in lottery models. Section 6.3 introduces lottery models and presents our main result (Theorem 6.3.2). In Section 6.4 we consider the case of neutral effectivity functions, for which a natural and simple lottery model can be based on the so-called uniform core. Section 6.5 concludes.

## 6.2 Nash consistent representation: an extension

In this section we assume that the set of alternatives is some finite set $Z$, containing at least two alternatives. Choices for $Z$ include our usual finite set of social states $A$, augmented with equal chance lotteries over $A$, to be introduced later.

Let $U$ be a non-empty set of *utility functions* $u : Z \to \mathbb{R}$, and let $X \in P_0(Z)$. Call $X$ *admissible with respect to* $U$ if there is a $u \in U$ such that $u(x) > u(y)$ for every $x \in X$ and every $y \in Z \setminus X$. Hence, a player with utility function $u$ strictly prefers every element of $X$ to every element not in $X$.

The following theorem extends Theorem 3.3.10. In this theorem the intersection condition (3.6) is weakened to condition (6.1) by making it conditional on admissibility of the involved sets.

**Theorem 6.2.1.** *Let $E : P(N) \to P(P_0(Z))$ be a superadditive and monotonic effectivity function. Then $E$ has a representation that is Nash consistent on $U^N$ if and only if*

$$[X^i \in E^*(i) \text{ and } X^i \text{ admissible w.r.t. } U \text{ for all } i \in N] \Rightarrow \bigcap_{i=1}^{n} X^i \neq \emptyset. \quad (6.1)$$

*Proof.* First assume that $E$ has a Nash consistent representation $\Gamma$ on $U^N$. For each $i \in N$, let $X^i \in E^*(i)$ be an admissible set and $u^i \in U$ such that $u^i(z) > u^i(y)$ for all $z \in X^i$ and $y \in Z \setminus X^i$. Let $x \in Z$ be a Nash equilibrium outcome of $(\Gamma, u^N)$. Then, by Proposition 3.2.1,

$$L(x, u^i) \in E^{\Gamma}(N \setminus \{i\}) = E(N \setminus \{i\}) \text{ for all } i \in N.$$

This implies $X^i \cap L(x, u^i) \neq \emptyset$ for every $i \in N$. By the choice of $u^i$, this implies $x \in X^i$ for every $i \in N$, hence $\bigcap_{i \in N} X^i \neq \emptyset$, so that (6.1) holds.

For the converse, assume (6.1). Let $u^N \in U^N$. For every $i \in N$ let $Y^i := \{y \in Z \mid Z \setminus L(y, u^i) \in E^*(i)\}$ and define

$$X^i = \begin{cases} \bigcap_{y \in Y^i} Z \setminus L(y, u^i) & \text{if } Y^i \neq \emptyset \\ Z & \text{otherwise.} \end{cases}$$

Then $X^i \in E^*(i)$ and $X^i$ is admissible with respect to $U$ for each $i \in N$, so by (6.1), $\bigcap_{i\in N} X^i \neq \emptyset$. Take $x \in \bigcap_{i\in N} X^i$. Then, by definition of $X^i$, $Z\backslash L(x,u^i) \notin E^*(i)$ for each $i \in N$. Hence, there is some set $Z^i \subseteq L(x,u^i)$ with $Z^i \in E(N\backslash\{i\}$, so by monotonicity of $E$, $L(x,u^i) \in E(N\backslash\{i\})$ for every $i \in N$. Theorem 3.2.3 now implies that $E$ has a Nash consistent representation. □

## 6.3 Lottery models

We shall now be more specific about the set $Z$. Let $A$ be a finite set of alternatives, $|A| \geq 2$. For each $B \in P_0(A)$, $\ell(B)$ denotes the lottery that assigns equal probability $1/|B|$ to each alternative in $B$. The set of all such equal chance lotteries with support in a set $B \in P_0(A)$ is denoted by $\widetilde{B}$, hence

$$\widetilde{B} = \{\ell(B') \mid B' \in P_0(B)\}.$$

By identifying each $x \in A$ with the degenerate lottery $\ell(\{x\})$, we have $B \subseteq \widetilde{B}$.

Typically, we shall consider the case $Z = \widetilde{A}$. Let $\widetilde{E} : P(N) \to P(P_0(\widetilde{A}))$ be an effectivity function. With $\widetilde{E}$ we associate an effectivity function $E : P(N) \to P(P_0(A))$ as follows. Let $E(\emptyset) = \emptyset$. For $S \in P_0(N)$ and $B \in P_0(A)$, we let $B \in E(S)$ if and only if there exists an $X \in \widetilde{E}(S)$ such that

$$B = \bigcup_{B' \in P_0(A):\ \ell(B') \in X} B'. \tag{6.2}$$

In other words, elements of $E(S)$ are obtained by taking the union of the supports of elements of $\widetilde{E}(S)$. It is straightforward to check that, indeed, $E$ is an effectivity function. If $E$ is derived from $\widetilde{E}$ in this way, then we call $\widetilde{E}$ a *lottery model for E*.

*Remark 6.3.1.* Monotonicity of $\widetilde{E}$ with respect to the players implies monotonicity of $E$ with respect to the players, and monotonicity of $\widetilde{E}$ with respect to the alternatives implies monotonicity of $E$ with respect to the alternatives. (To show the latter claim, if $B \in E(S)$ resulting from $\widetilde{B} \in \widetilde{E}(S)$, and $C \supseteq B$, then $C \in E(S)$ follows from considering $\widetilde{B} \cup C \in \widetilde{E}(S)$. Note that $A \subseteq \widetilde{A}$.) Hence, monotonicity of $\widetilde{E}$ is inherited by $E$. If $\widetilde{E}$ is monotonic and superadditive, then also $E$ is superadditive. For let $\widetilde{E}$ be monotonic and superadditive, $S_1, S_2 \in P_0(N)$ with $S_1 \cap S_2 = \emptyset$, and let $B_1 \in E(S_1)$ and $B_2 \in E(S_2)$. Let $X_1 \in \widetilde{E}(S_1)$ and $X_2 \in \widetilde{E}(S_2)$ correspond to $B_1$ and $B_2$ as in the definition of $E$, i.e., as in (6.2). Then superadditivity of $\widetilde{E}$ implies $X := X_1 \cap X_2 \in \widetilde{E}(S_1 \cup S_2)$, hence

$$E(S_1 \cup S_2) \ni \bigcup_{B' \in P_0(A): \ell(B') \in X} B' \subseteq B_1 \cap B_2.$$

Monotonicity of $E$ now implies $B_1 \cap B_2 \in E(S_1 \cup S_2)$. This shows that $E$ is superadditive. The converse is not true: a lottery model $\widetilde{E}$ for a monotonic and superadditive effectivity function $E$ is not itself necessarily monotonic and superadditive.

Let $u : \widetilde{A} \to \mathbb{R}$ and suppose that $u(x_1) \geq u(x_2) \geq \ldots \geq u(x_m)$, where $A = \{x_1, x_2, \ldots, x_m\}$. For $\ell \in \widetilde{A}$ and $i \in \{1, 2, \ldots, m\}$ let $\ell_i$ be the probability assigned by $\ell$ to $x_i$. We say that $u$ *respects stochastic dominance* if $u(\ell) \geq u(\ell')$ whenever $\ell, \ell' \in \widetilde{A}$ satisfy

$$\sum_{i=k}^{m} \ell_i \leq \sum_{i=k}^{m} \ell'_i \text{ for all } k = 1, 2, \ldots, m.$$

We assume that lotteries are evaluated by utility functions satisfying this condition. Therefore, we define

$$U_{\text{sd}} := \left\{ u \in \mathbb{R}^{\widetilde{A}} \mid u \text{ respects stochastic dominance} \right\}.$$

The set $U_{\text{sd}}$ contains in particular the set of expected utility functions

$$\left\{ u \in \mathbb{R}^{\widetilde{A}} \mid u(\ell(B)) = \sum_{a \in B} \frac{u(a)}{|B|} \text{ for all } B \in P_0(A) \right\}.$$

The main result of this chapter is that for every monotonic and superadditive effectivity function there exists a lottery model which has a Nash consistent representation on $U_{\text{sd}}^N$. Clearly, monotonicity and superadditivity cannot be left out here: a lottery model that has a representing game form must be monotonic and superadditive, and by Remark 6.3.1 the original 'deterministic' effectivity function must also be monotonic and superadditive. But, in contrast to Theorem 6.2.1, no additional condition is needed on $E$.

**Theorem 6.3.2.** *Let $E : P(N) \to P(P_0(A))$ be a monotonic and superadditive effectivity function. Then there exists an effectivity function $\widetilde{E} : P(N) \to P(P_0(\widetilde{A}))$ such that*

*(i) $\widetilde{E}$ is a lottery model for $E$;*
*(ii) $\widetilde{E}$ has a representation which is Nash consistent on $U_{\text{sd}}^N$.*

*Proof.* Define $\widetilde{E} : P(N) \to P(P_0(\widetilde{A}))$ as follows. Let $\widetilde{E}(\emptyset) = \emptyset$ and for every $S \in P_0(N)$ with $|S| \neq n - 1$ let

$$\widetilde{E}(S) = \left\{ X \in P_0(\widetilde{A}) \mid X \supseteq \widetilde{B} \text{ for some } B \in E(S) \right\}.$$

In order to define $\widetilde{E}$ for $(n-1)$-person coalitions we introduce a notation. For each $i \in N$, $C \in E(N \setminus \{i\})$ and $C' \in P_0(C)$ define the set $X(C, C') \in P_0(\widetilde{A})$ by

$$X(C, C') = \begin{cases} \{\ell(\{c\} \cup C') \mid c \in C \setminus C'\} & \text{if } C' \neq C \\ \{\ell(C)\} & \text{if } C' = C. \end{cases}$$

Note in particular that the union of the supports of the elements of the set $X(C, C')$ is equal to $C$.

Let now $S \in P_0(N)$ with $|S| = n - 1$, say $S = N \setminus \{i\}$ for some $i \in N$. Then we define $X \in \widetilde{E}(N \setminus \{i\})$ if and only if $X \supseteq \widetilde{B}$ for some $B \in E(N \setminus \{i\})$ or $X \supseteq X(C, C') \cup \widetilde{B}$ for some $C \in E(N \setminus \{i\})$, $C' \in P_0(C)$, and $B \in P_0(A)$ such that $B \cap B' \neq \emptyset$ for all $B' \in E(\{i\}$. This concludes the definition of $\widetilde{E}$. It is straightforward to verify that $\widetilde{E} : P(N) \to P(P_0(\widetilde{A}))$ is a monotonic and superadditive EF and that $\widetilde{E}$ is a lottery model for $E$.

It remains to prove that $\widetilde{E}$ has a Nash consistent representation on $U_{\text{sd}}^N$. For each $i \in N$, let $X^i \in \widetilde{E}^*(i)$ such that $X^i$ is admissible with respect to $U_{\text{sd}}$. In view of Theorem 6.2.1 it is sufficient to prove $\bigcap_{i \in N} X^i \neq \emptyset$. For each $i \in N$ choose $u^i \in U_{\text{sd}}$ such that

$$u^i(x) > u^i(y) \text{ for all } x \in X^i \text{ and } y \in \widetilde{A} \setminus X^i \tag{6.3}$$

(this is possible since each $X^i$ is admissible), and choose $B^i \in E(\{i\})$ and $b^i \in B^i$ such that both

$$u^i(b^i) = \min\{u^i(b) \mid b \in B^i\} \geq \min\{u^i(b) \mid b \in B\} \text{ for all } B \in E(\{i\})$$

and

$$B^i = \{b \in A \mid u^i(b^i) \leq u^i(b)\}.$$

(This is possible in view of monotonicity of $E$.)

Also, for each $i \in N$, define $C^i := \bigcap_{j \in N \setminus \{i\}} B^j$. Then $C^i \in E(N \setminus \{i\})$ by superadditivity of $E$. Choose $a^i \in C^i$ such that

$$u^i(a^i) = \max\{u^i(a) \mid a \in C^i\}.$$

Since, by superadditivity of $E$, $C^i \cap B^i \neq \emptyset$, we have $u^i(a^i) \geq u^i(b^i)$.

Now fix a player $i \in N$ and write $u^i(x_1) \geq u^i(x_2) \geq \ldots \geq u^i(x_m)$, where $A = \{x_1, x_2, \ldots, x_m\}$. Let $k \in \{1, 2, \ldots, m\}$ such that $a^i = x_k$. Choose $p$ with $k \leq p \leq m$ such that

$$u^i(\ell(\{x_k, x_p, x_{p+1}, \ldots, x_m\})) \leq u^i(\ell(\{x_k, x_{p'}, x_{p'+1}, \ldots, x_m\}))$$

for all $k \leq p' \leq m$. Consider the set $D = \{x_k, \ldots, x_m\}$. Since $C^i \subseteq D$, we have $D \in E(N \setminus \{i\})$. Let $D' := \{x_p, x_{p+1}, \ldots, x_m\} \subseteq D$. Define $Y \in P_0(\widetilde{A})$ by

$$Y = X(D, D') \cup \widetilde{F}$$

where

$$F = \left\{b \in A \mid u^i(b^i) \geq u^i(b)\right\}.$$

Observe that, by definition of $b^i$, we have $F \cap B' \neq \emptyset$ for every $B' \in E(\{i\})$. By definition of $\widetilde{E}$ and $Y$, we have $Y \in \widetilde{E}(N \setminus \{i\})$. It follows, in particular, that the set $X^i$ contains an element of $Y$, say $y$. Consider the lottery $\bar{\ell} = \ell(\bigcap_{j \in N} B^j)$. Then

$$a^i \in B^i \cap C^i = \bigcap_{j\in N} B^j \text{ and } u^i(a^i) \geq u^i(c) \geq u^i(b^i) \text{ for all } c \in \bigcap_{j\in N} B^j. \tag{6.4}$$

We show that $\bar{\ell} \in X^i$ by considering all the possible values for $y \in Y \cap X^i$.

If $y \in \widetilde{F}$, then $u^i(\bar{\ell}) \geq u^i(b^i) \geq u^i(y)$, where the first inequality follows from (6.4) and the last inequality by definition of $F$ and the fact that $u^i$ respects stochastic dominance. By (6.3), this implies $\bar{\ell} \in X^i$.

If $y \in X(D, D')$, then $y = \ell(\{x_{p'}, x_p, \dots, x_m\})$ for some $p' \in \{k, k+1, \dots, p-1\}$ if $k < m$ and $y = x_m$ if $k = m$. In that case, we argue as follows. Write $\bigcap_{j\in N} B^j = \{a^i, y_1, \dots, y_r\}$ with $u^i(a^i) \geq u^i(y_1) \geq u^i(y_2) \geq \dots \geq u^i(y_r)$. Then

$$\begin{aligned} u^i(\bar{\ell}) &\geq u^i(\ell(\{a^i, x_{m-r+1}, x_{m-r+2}, \dots, x_m\})) \\ &\geq u^i(\ell(\{a^i, x_p, \dots, x_m\})) \\ &\geq u^i(\ell(\{x_{p'}, x_p, \dots, x_m\})) \\ &= u^i(y) \end{aligned}$$

where the second inequality follows from the choice of $p$, and the third follows since $u^i(a^i) \geq u^i(x_{p'})$. (Note that for the first and third inequalities the fact that $u^i$ respects stochastic dominance is used.) Hence also in this case, (6.3) implies that $\bar{\ell} \in X^i$.

Since $i \in N$ was arbitrary, we conclude that $\bar{\ell} \in X^j$ for every $j \in N$, hence $\bigcap_{j\in N} X^j \neq \emptyset$. □

The following example illustrates the effectivity function $\widetilde{E}$, constructed in the proof of Theorem 6.3.2, for the effectivity function associated with a $2 \times 2$ bimatrix game form.

*Example 6.3.3.* Let $N = \{1, 2\}$, $A = \{a, b, c, d\}$, and consider the effectivity function $E$ derived from the game form

$$\begin{pmatrix} a & b \\ c & d \end{pmatrix}$$

where player 1 chooses a row and player 2 chooses a column. In particular, $E(\{1\})$ contains $\{a, b\}$, $\{c, d\}$, and all supersets; $E(\{2\})$ contains $\{a, c\}$, $\{b, d\}$, and all supersets. It is easy to see that condition (6.1) in Theorem 6.2.1 (with $Z = A$) is not fulfilled. For instance, $\{a, d\} \in E^*(\{1\})$, $\{b, c\} \in E^*(\{2\})$, both are (trivially) admissible with respect to $\mathbb{R}^A$, but $\{a, d\} \cap \{b, c\} = \emptyset$. [We recall the explanation (cf. Example 3.3.11) of why a Nash consistent representation cannot exist. Observe that $\{a, d\} \in E^*(\{1\})$ means that for any given strategy of player 2 in some representing game form, player 1 can make sure that the final outcome is in $\{a, d\}$. A similar statement holds for player 2 and the set $\{b, c\}$. Now suppose that preferences are such that player 1 prefers $a$ and $d$ over $b$ and $c$ and player 2 prefers $b$ and $c$ over $a$ and $d$: then, clearly, a Nash equilibrium cannot exist. This illustrates, again, the necessity of the intersection condition (6.1).]

The effectivity function $\widetilde{E}$, constructed in the first two paragraphs of the proof of Theorem 6.3.2, assigns the following sets (where, e.g., $ab$ is shorthand for $\ell(\{a,b\})$, the equal chance lottery on $\{a,b\}$):

$$\begin{aligned}\widetilde{E}(\{1\}):&\ \{a,b,ab\}\ \{a,d,ab\}\ \{c,b,ab\}\ \{c,d,ab\}\\ &\ \{a,b,cd\}\ \{a,d,cd\}\ \{c,b,cd\}\ \{c,d,cd\}\\ \widetilde{E}(\{2\}):&\ \{a,c,ac\}\ \{a,d,ac\}\ \{b,c,ac\}\ \{b,d,ac\}\\ &\ \{a,c,bd\}\ \{a,d,bd\}\ \{b,c,bd\}\ \{b,d,bd\}\end{aligned}$$

and all supersets of these sets within $\widetilde{A}$. It is easy to check that $\widetilde{E}$ is a lottery model for $E$. By the proof of Theorem 6.3.2, $\widetilde{E}$ has a Nash consistent representation. This can also be verified 'directly' by using Theorem 6.2.1 and showing that $X^1 \cap X^2 \neq \emptyset$ for all *admissible* $X^1 \in \widetilde{E}^*(\{1\})$ and $X^2 \in \widetilde{E}^*(\{2\})$, but this is a rather tedious task. Admissibility has strong implications. For instance, if $a, b \in X^1$, then also $ab \in X^1$, or if $ab \in X^1$, then also $a \in X^1$ or $b \in X^1$, etc.

The Gibbard Paradox (Example 3.3.11) is an instance of this example. Theorem 6.3.2 shows how it can be resolved by allowing lotteries, in the way as described above.

For particular cases, there may exist other lottery models that are less complex than the one constructed in the proof of Theorem 6.3.2 and in that sense more attractive. This is the case in the next example, where the unanimity effectivity function is considered.

*Example 6.3.4.* Let $E : P(N) \to P(P_0(A))$ be the *unanimity effectivity function*, i.e., $E(S) = \{A\}$ for all $S \in P_0(N)$, $S \neq N$. This effectivity function clearly fails to satisfy condition (6.1). It can be checked that here the lottery $\bar{\ell}$ in the proof of Theorem 6.3.2 is equal to $\ell(A)$, but in this case the lottery model $\widetilde{E}$ in that proof is overly complicated. It is straightforward to see that also the effectivity function $\widetilde{E}'$ is a lottery model for $E$, where for each $S \in P_0(N)$, $S \neq N$, $\widetilde{E}'(S)$ consists of $\{\ell(A)\}$ and all its supersets in $\widetilde{A}$, and $\widetilde{E}'(N) = P_0(\widetilde{A})$. The effectivity function $\widetilde{E}'$ is different from but simpler than $\widetilde{E}$. By applying Theorem 6.2.1 and checking condition (6.1) – for each player $i$ each element of $\widetilde{E}^*(\{i\})$ must contain $\ell(A)$ – it follows that this lottery model has a Nash consistent representation.

Example 6.3.4 is a special case of a neutral effectivity function. These effectivity functions are studied in the next section.

## 6.4 Neutral effectivity functions

For convenience we recall from Section 3.5.1 some facts about veto functions. A veto function is a function $v : P(N) \to \{-1, 0, \ldots, |A| - 1\}$ such that

$v(\emptyset) = -1$, $v(S) \geq 0$ if $S \in P_0(N)$, and $v(N) = |A| - 1$. The interpretation is that coalition $S$ can veto any subset of the alternatives with at most $v(S)$ elements. With $v$ we can associate a neutral (i.e., not depending on the names of the alternatives) effectivity function $E_v$ by

$$E_v(S) = \{B \in P_0(A) \mid v(S) \geq |A \setminus B|\} = \{B \in P_0(A) \mid v(S) \geq |A| - |B|)\}$$

for every $S \in P(N)$. Conversely, every neutral effectivity function is derived from some veto function. A veto function is monotonic if $v(S) \leq v(S^*)$ for all $S, S^*$ with $S \subseteq S^*$, and superadditive if $v(S) + v(S^*) \leq v(S \cup S^*)$ for all $S, S^* \in P(N)$ with $S \cap S^* = \emptyset$. A veto function is monotonic [superadditive] if and only if the associated effectivity function is monotonic [superaditive].

We shall show that for neutral effectivity functions there exists a simple and quite natural lottery model that has a Nash consistent representation. To this end we need the concept of the uniform core.

Let $E : P(N) \to P(P_0(A))$ be a monotonic and superadditive effectivity function. Let $U = \mathbb{R}^A$, $u^N \in U^N$, and say that $x \in A$ is *uniformly dominated* by $B \in P_0(A)$ via $S \in P_0(N)$ if (i) $B \in E(S)$; (ii) $x \notin B$; and (iii) $u^i(b) > u^i(a)$ for all $b \in B$, $a \in A \setminus B$, and $i \in S$. We also say that $S$ *blocks* $x$ by $B$.

Observe that, if $x$ is uniformly dominated by $B$ via $S$, then $x$ is also dominated (cf. Definition 5.2.5) by $B$ via $S$. The converse is not true: for uniform domination we need that for every player in $S$ the set of alternatives better than $x$ is exactly the set $B$, for domination we only need that it contains $B$.

The set of all alternatives that are not uniformly dominated by some set $B$ via some coalition $S$ is called the *uniform core* and denoted $C_{\text{uf}}(E, u^N)$. Obviously, by the above, the core $C(E, u^N)$ is a subset of the uniform core $C_{\text{uf}}(E, u^N)$. While the core can be empty, the uniform core is never empty. This is proved in Abdou and Keiding (1991, Lemma 3.2). For completeness' sake we present a proof here.

**Lemma 6.4.1.** *Let $E : P(N) \to P(P_0(A))$ be a monotonic and superadditive effectivity function, and let $u^N \in U^N$. Then $C_{\text{uf}}(E, u^N) \neq \emptyset$.*

*Proof.* Suppose, to the contrary, that $C_{\text{uf}}(E, u^N) = \emptyset$. Write $A = \{x_1, \ldots, x_m\}$ and for each $j = 1, \ldots, m$ let $S_j \in P_0(N)$ and $B_j \in E(S_j)$ such that $x_j$ is uniformly dominated by $B_j$ via $S_j$. Then, since for each $j$ we have $x_j \notin B_j$, it follows that $\bigcap_{j=1}^m B_j = \emptyset$.

Now, without loss of generality, let $\{B_1, \ldots, B_r\}$, where $1 \leq r \leq m$, be those sets in $\{B_1, \ldots, B_m\}$ that are minimal under inclusion, where we take only one of two equal sets if any. We claim that the corresponding sets $S_1, \ldots, S_r$ are pairwise disjoint. Indeed, suppose for instance that $i \in S_1 \cap S_2$, and, say, $u^i(x_1) \geq u^i(x_2)$. Then, by definition of uniform domination it follows that $B_1 \subseteq B_2$, contradicting inclusion minimality of the sets in $\{B_1, \ldots, B_r\}$. By superadditivity we have $\bigcap_{j=1}^r B_j \in E(\bigcup_{j=1}^r S_j)$. Since, clearly, $\bigcap_{j=1}^m B_j = \bigcap_{j=1}^r B_j$, we obtain

$$\emptyset = \bigcap_{j=1}^{m} B_j = \bigcap_{j=1}^{r} B_j \in E(\bigcup_{j=1}^{r} S_j) ,$$

a contradiction. □

The uniform core represents $E$ in the following sense. If $S$ is effective for a set of alternatives $B$, then $S$ has a utility profile such that the associated uniform core is a subset of $B$ for every utility profile of the players outside $S$. Formally, we have the following lemma (cf. Keiding and Peleg, 2006a).

**Lemma 6.4.2.** *Let $E : P(N) \to P(P_0(A))$ be a monotonic and superadditive effectivity function. Then, for every $S \in P_0(N)$ and every $B \in P_0(A)$,*

$$B \in E(S) \Leftrightarrow \exists u^S \in U^S \; \forall u^{N\setminus S} \in U^{N\setminus S} : C_{uf}(E, (u^S, u^{N\setminus S})) \subseteq B.$$

*Proof.* Let $S \in P_0(N)$ and $B \in P_0(A)$.

First, suppose $B \in E(S)$. For each $i \in S$ let $u^i \in U$ be defined by $u^i(x) = 1$ for all $x \in B$ and $u^i(x) = 0$ for all $x \in A \setminus B$. Then all $x \in A \setminus B$ are blocked by $S$ using $B$, so that $C_{uf}(E, (u^S, u^{N\setminus S})) \subseteq B$ for all $u^{N\setminus S} \in U^{N\setminus S}$.

Second, for the converse, let $u^S \in U^S$ such that $C_{uf}(E, (u^S, u^{N\setminus S})) \subseteq B$ for all $u^{N\setminus S} \in U^{N\setminus S}$. First observe that, if $x \in A \setminus B$, then $x$ must be blocked via some coalition $S' \subseteq S$. Indeed, otherwise take, for each $i \notin S$, a preference $u^i$ with $u^i(x) = 1$ and $u^i(y) = 0$ for all $y \neq x$: then no player outside $S$ can participate in blocking $x$, so $x \in C_{uf}(E, (u^S, u^{N\setminus S})) \subseteq B$, a contradiction.

Now, for each player $i \in S$, take $x^i \in A$, $S^i \subseteq S$ with $i \in S^i$, and $B^i \in E(S^i)$ such that (i) $x^i$ is blocked by $B^i$ via $S^i$ and (ii) for each $y \in B^i$, $y$ is not blocked via any coalition $S' \subseteq S$ with $i \in S'$; if such a triple does not exist for some player $i$, then we take $B^i$ equal to $A$. Without loss of generality let $S = \{1, \ldots, |S|\}$ and let $\{B^1, \ldots, B^k\}$ with $1 \leq k \leq |S|$ be the subset of those elements of $\{B^1, \ldots, B^{|S|}\}$ that are minimal under inclusion and all different (if two minimal sets $B^i$ and $B^j$ are equal then take only one of the two). By the same argument as in the proof of Lemma 6.4.1 the associated coalitions $S^1, \ldots, S^k$ are pairwise disjoint and thus, by superadditivity and monotonicity,

$$\bigcap_{i=1}^{|S|} B^i = \bigcap_{i=1}^{k} B^i \in E(\bigcup_{i=1}^{k} S^i) \subseteq E(S) .$$

Consider any $x \in \bigcap_{i=1}^{|S|} B^i$. If $x$ were blocked by some coalition $S' \subseteq S$, then this would violate condition (ii) in the definition of the triple $x^i, S^i, B^i$ for the players $i \in S'$, a contradiction. By the argument in the second paragraph of the proof, it follows that $x \in B$. Hence, $\bigcap_{i=1}^{|S|} B^i \subseteq B$, so that $B \in E(S)$ by monotonicity. □

We can now construct a lottery model for $E$ based on the uniform core. We formulate this as a lemma: the proof is straightforward using Lemma 6.4.2.

**Lemma 6.4.3.** *Let $E : P(N) \to P(P_0(A))$ be a monotonic and superadditive effectivity function. Define $\widetilde{E}_{uf} : P(N) \to P(P_0(\widetilde{A}))$ by requiring for each $S \in P_0(N)$ and $X \in P_0(\widetilde{A})$:*

$$X \in \widetilde{E}_{uf}(S) \Leftrightarrow \exists u^S \in U^S \ \forall u^{N\setminus S} \in U^{N\setminus S} : \ell\left(C_{uf}(E,(u^S,u^{N\setminus S}))\right) \in X.$$

*Then $\widetilde{E}_{uf}$ is a monotonic and superadditive lottery model for $E$.*

The construction of $\widetilde{E}_{\text{uf}}$ implies, in fact, that it is the effectivity function associated with the game form $\Gamma_{\text{uf}} = (U,\ldots,U;g;\widetilde{A})$, defined by $g(u^N) = \ell(C_{\text{uf}}(E,u^N))$ for each $u^N \in U^N$. Hence, the game form $\Gamma_{\text{uf}}$ represents the effectivity function $\widetilde{E}_{\text{uf}}$, which in turn is a lottery model for $E$. We will show that $\Gamma_{\text{uf}}$ is Nash consistent. Observe that $\Gamma_{\text{uf}}$ is in fact a social choice function, where each player just reports a weak ordering over the alternatives in the form of a utility function – in fact, it is crucial for the proof of Theorem 6.4.4 below that the reported ordering can be weak. Given such a profile of reports one computes the uniform core and the outcome of the game is the equal chance lottery over the elements of the uniform core.

**Theorem 6.4.4.** *Let $u^N \in U^N_{\text{sd}}$. Then the game $(\Gamma_{uf}, u^N)$ has a Nash equilibrium.*

*Proof.* We construct a strategy profile $\hat{u}^N \in U^N$ inductively as follows. First, let $W(1) \subseteq A$ contain exactly $v(1)$ worst alternatives according to $u^1$, that is, $u^1(x) \leq u^1(y)$ for all $x \in W(1)$ and $y \in A \setminus W(1)$. Define $\hat{u}^1(x) = 0$ and $\hat{u}^1(y) = 1$ for all $x \in W(1)$ and $y \in A\setminus W(1)$. Let $k \in \{2,\ldots,n\}$ and suppose that $\hat{u}^l$ has been defined for all $1 \leq l \leq k-1$. Then let $W(k) \subseteq A$ contain exactly $v(k)$ worst alternatives in $A \setminus \bigcup_{l=1}^{k-1} W(l)$ according to $u^k$, and define $\hat{u}^k(x) = 0$ and $\hat{u}^k(y) = 1$ for all $x \in W(k)$ and $y \in A \setminus W(k)$.

We claim that $\hat{u}^N$ is a Nash equilibrium in $(\Gamma_{\text{uf}}, u^N)$. Let $k \in N$ and assume that each player $l \in N \setminus \{k\}$ plays the strategy $\hat{u}^l$. Consider any coalition $S \subseteq N\setminus\{k\}$ with more than one player. Then, because of the strict inequality sign in condition (iii) of the definition of uniform domination, $S$ could only possibly block some alternative by the set $A\setminus W(l)$ for some $l \in S$, but all these sets are different since all sets $W(l)$ are different. Hence, only singletons in $N\setminus\{k\}$ block: each $l \in N\setminus\{k\}$ blocks $W(l)$, so altogether the set $\bigcup_{l\in N\setminus\{k\}} W(l)$ is blocked by the single players in $N\setminus\{k\}$. Consider the decision problem for player $k$. By the same argument as before, a non-singleton coalition $S$ containing player $k$ can only possibly block some alternative if $S = \{k,j\}$ for some $j \neq k$ (since all sets $W(l)$, $l \in N\setminus\{k\}$ are different), but in that way $S$ can only block the set $W(j)$, namely by player $k$ playing some strategy $u'$ such that $u'(x) < u'(y)$ for all $x \in W(j)$ and $y \in A\setminus W(j)$. Then the game would result in the equal chance lottery $\ell\left(A \setminus \bigcup_{l\in N\setminus\{k\}} W(l)\right)$. By only using his own blocking power, however, player $k$ can make sure that the outcome of the game is $\ell\left(A \setminus \bigcup_{l\in N} W(l)\right)$ by playing $\hat{u}^k$. Since player $k$'s utility function $u^k$ respects stochastic dominance, this is clearly an improvement for player $k$, and also the best outcome attainable by using $k$'s own blocking power. □

The Nash equilibrium exhibited in the proof of Theorem 6.4.4 is a very natural one, since it consists of successive sincere vetoing of alternatives: first, player 1 vetoes his $v(1)$ worst alternatives, next, player 2 vetoes his $v(2)$ worst alternatives of the remaining ones, etc. Of course, vetoing according to any other ordering of the players would also be a Nash equilibrium. These specific equilibria have the drawback that they need not be Pareto optimal. For instance, if all players have the same preference, with a unique top alternative, but $\sum_{i \in N} v(i) < |A| - 1$, then the resulting lottery does not put probability 1 on the common top alternative. Of course, in this example the profile in which every player reports his true preference is also a Nash equilibrium: the uniform core associated with this profile consists of the common top alternative and, thus, the degenerate lottery that puts all probability on this alternative results.

If $E$ is non-neutral, then $\widetilde{E}_{\mathrm{uf}}$ is still a lottery model for $E$ and $\widetilde{E}_{\mathrm{uf}} = E^{\Gamma_{\mathrm{uf}}}$, but it is not clear whether $\Gamma_{\mathrm{uf}}$ is still Nash consistent.

## 6.5 Notes and comments

In this chapter, which is based on Peleg and Peters (2009), we have proved that every (monotonic and superadditive) effectivity function can be augmented, by adding finitely many equal chance lotteries, to a new effectivity function (lottery model) which preserves the original effectivity and has a Nash consistent representation. This approach is based on two particular assumptions. We elaborate on these assumptions in the next two remarks.

*Remark 6.5.1.* First, we assume that in the lottery model the original effectiveness of a coalition $S$ of players for a set $B$ of alternatives is preserved if $S$ is now effective for some set $X$ of equal chance lotteries such that the union of the supports of the lotteries in $X$ is equal to $B$. For instance, if $B = \{a, b, c\}$, then $X$ could be the one-point set $\{\ell(\{a, b, c\})\}$ but also the two-point set $\{a, \ell(\{a, b, c\})\}$. This example shows that, in this set-up, we cannot really interpret effectiveness for $B$ as the alternatives of $B$ being equiprobable, even if we only add equal chance lotteries. Rather, players (or coalitions) evaluate effectiveness purely in terms of supports.

*Remark 6.5.2.* Second, we assume that equal chance lotteries resulting as outcomes of the representing game form are evaluated by utility functions respecting first order stochastic dominance. This is a minimal requirement and therefore hardly controversial. A special case of this is expected utility. For a justification of the use of expected utility see Fishburn (1972), where preferences on sets of alternatives are considered and the expected utility property for equal chance lotteries is derived from conditions on these preferences. The assumption of equal chance lotteries evaluated by expected utility has been made frequently in the social choice literature, such as in Barberà, Dutta, and Sen (2001), but also in earlier work, e.g., Feldman (1980). In these works, outcomes can be sets, which are evaluated as equal chance lotteries using expected utility. In fact, this was also done in Section 6.4, where we considered the uniform core and evaluated that set as an equal chance lottery.

The assumption of utility functions respecting first order stochastic dominance is called 'monotonicity' in Abreu and Sen (1991).

We next comment on Pareto optimality in relation to lottery models.

*Remark 6.5.3.* By using the game form $\widetilde{\Gamma}_0$, constructed in the proof of Theorem 2.4.7, to represent a lottery model $\widetilde{E}$, we obtain again weak acceptability: for any profile of preferences there there is a Nash equilibrium with Pareto optimal outcome. This follows since Theorem 3.3.13 continues to hold in the extended framework of Theorem 6.2.1.

What does this mean? Suppose $u^N \in U^N_{\text{sd}}$ is a profile of preferences and $a, b \in A$ such that $u^i(a) > u^i(b)$ for all $i \in N$. Then, clearly, a lottery that has $b$ but not $a$ in its support is not Pareto optimal, and so there is a Nash equilibrium where this lottery is not the associated outcome. On the other hand, it is not difficult to come up with an example of a Pareto optimal lottery containing both $a$ and $b$, since we only allow equal-chance lotteries.[1] Thus, Pareto optimality of a lottery does not imply that only Pareto optimal pure alternatives occur in the support.

Our final comment is related to the avoidance of mixed strategies in the game form associated with a lottery model.

*Remark 6.5.4.* The main result in this chapter is also a contribution to the classical 'purification' problem – e.g., Harsanyi (1973). For any finite game form, it enables us to construct a new finite game form which preserves the strategic possibilities of players and coalitions in the sense that the associated effectivity function is a lottery model for the effectivity function associated with the original game form, and which has a pure Nash equilibrium for any profile of utility functions respecting first order stochastic dominance among equal chance lotteries.

[1] E.g., $N = \{1, 2\}$, $A = \{a, b, c\}$, $u^1(c) = 2$, $u^1(a) = 0.4$, $u^1(b) = 0$, $u^2(a) = 2$, $u^2(b) = 1.3$, $u^2(c) = 0$; assume expected utility. Then $\ell(A)$ is Pareto optimal, although both agents strictly prefer $a$ to $b$.

# Chapter 7
# On the continuity of representations of constitutions

## 7.1 Motivation and summary

In the previous chapters we have disregarded the topological properties of the strategy sets and the outcome functions of representations of topological effectivity functions. In this chapter we enquire about the existence of representations of which the strategy sets and outcomes space are compact metric spaces and where the outcome function is continuous. Clearly, the continuity of the outcome function of a representation in the strategies played by the members of the society is very desirable. Unfortunately, in Section 7.2 we describe an effectivity function which admits no continuous representation. Fortunately – and this is a main message of this chapter – continuity properties of representations of topological effectivity functions are latent everywhere in our model.

In Section 7.3 we prove that every (monotonic and superadditive) effectivity function that is generated by a finite set of closed subsets of alternatives has a continuous representation. This leads to the result that effectivity functions with a continuous representation are dense in the set of all topological effectivity functions.

Let $\mathcal{C}$ be the Cantor (ternary) set. It is a classical result that for every compact metric space $A$ there exists a continuous surjection $f_A : \mathcal{C} \to A$. In Section 7.4 we observe that for all such $A$ and $f_A$, every effectivity function $E$ on $A$ can be 'lifted' to an effectivity function $\widetilde{E}$ on $\mathcal{C}$. Furthermore, every [continuous] representation of $\widetilde{E}$ yields naturally a [continuous] representation of $E$. Thus, using the mentioned classical result we show that the general representation problem can be reduced to the representation problem on $\mathcal{C}$.

Section 7.5 contains the following important result: on $\mathcal{C}$ (or, as a matter of fact, on any compact subset of the real line) every effectivity function with closed values has an upper (or lower) semicontinuous representation. Semicontinuity is a weaker form of continuity that nevertheless allows a generalization of the Weierstrass Theorem: an upper [lower] semicontinuous function

B. Peleg and H. Peters, *Strategic Social Choice*, Studies in Choice and Welfare,
DOI 10.1007/978-3-642-13875-1_7, 

on a compact set attains a maximum [minimum]. We also show that this semicontinuous representation is Nash consistent.

Using the techniques of Section 7.4 we show in Section 7.6 that the result in Section 7.5 implies that every effectivity function (over an arbitrary compact metric space) has a representation of which the outcome function is a modified Baire function of class 2. This is, again, a weaker continuity property.

## 7.2 Continuous representations may not exist

In this section we provide an example of an effectivity function that does not have a continuous representation. Let $A$ be a compact metric space and let $E : P(N) \to P(\mathcal{K}(A))$ be an effectivity function – this will be the setting throughout the chapter. (Recall that $\mathcal{K}(A)$ denotes the set of all non-empty closed subsets of $A$.) A game form $\Gamma = (N; \Sigma^1, \ldots, \Sigma^n; g; A)$ is a *continuous* representation of $E$ if (i) $\Gamma$ is a representation of $E$; and (ii) $\Sigma^1, \ldots, \Sigma^n$ are compact metric spaces and $g : \Sigma = \Sigma^1 \times \ldots \times \Sigma^n \to A$ is continuous when $\Sigma$ is endowed with the product topology.

REMARK If $\bar{d}^i$ is a metric on $\Sigma^i$ for each $i \in N$, then in order to obtain the product topology one may use the metric $\bar{d} = \sum_{i=1}^n \bar{d}^i$ on $\Sigma$, i.e., $\bar{d}\left((\sigma^1, \ldots, \sigma^n), (\mu^1, \ldots, \mu^n)\right) = \sum_{i=1}^n \bar{d}^i(\sigma^i, \mu^i)$.

When $A$ and all $\Sigma^i$ are finite then in the discrete topology every representation is continuous.

We shall now give an example of a compact metric space $A$ of alternatives and a topological effectivity function $E$ which has no continuous representation.

*Example 7.2.1.* Let

$$A = \{(x, y) \in \mathbb{R}^2 \mid x \geq 0, y \geq 0, \text{ and } x + y \leq 1\} .$$

Further, let $N = \{1, 2\}$,

$$E(1) = \{B \in \mathcal{K}(A) \mid B \supseteq [(1,0), (0,y)] \text{ for some } 0 \leq y \leq 1\} ,$$

and

$$E(2) = \{B \in \mathcal{K}(A) \mid B \supseteq [(0,1), (x,0)] \text{ for some } 0 \leq x \leq 1\} .$$

With $E(\emptyset) = \emptyset$ and $E(N) = \mathcal{K}(A)$, $E$ is a monotonic and superadditive effectivity function. See Figure 7.1 for an illustration.
We claim that $E$ has no continuous representation. Suppose, on the contrary, that $\Gamma = (N; \Sigma^1, \Sigma^2; g; A)$ is a continuous representation of $E$.
Let $\widetilde{y}(k) \uparrow 1$. For each $k$ there exists $\widetilde{\sigma}^1(k) \in \Sigma^1$ such that

$$g(\widetilde{\sigma}^1(k), \Sigma^2) = \{g(\widetilde{\sigma}^1(k), \sigma^2) \mid \sigma^2 \in \Sigma^2\} = [(1,0), (0, \widetilde{y}(k))] . \tag{7.1}$$

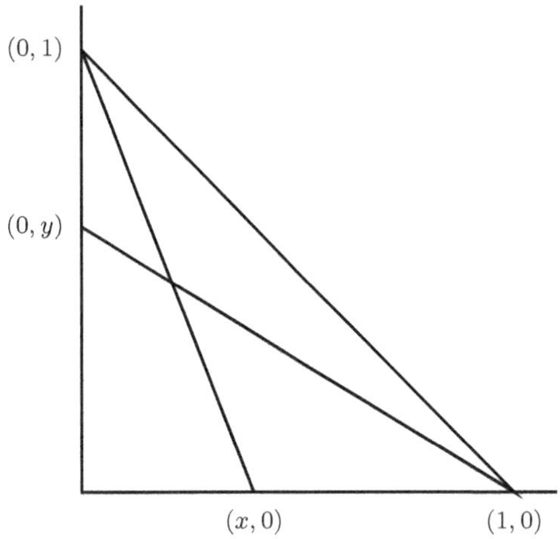


**Fig. 7.1** Example 7.2.1.

Let $\widetilde{\sigma}^1(k_j)$ with $\widetilde{\sigma}^1(k_j) \to \sigma_0^1$ for $j \to \infty$ be a converging subsequence. Then

$$g(\sigma_0^1, \Sigma^2) = [(1,0),(0,1)] \,. \tag{7.2}$$

Denote $\sigma^1(j) = \widetilde{\sigma}^1(k_j)$ and $y(j) = \widetilde{y}(k_j)$, $j = 1, 2, \ldots$
Let now $\widetilde{x}(k) \uparrow 1$. For each $k$ there exists $\widetilde{\sigma}^2(k) \in \Sigma^2$ such that

$$g(\Sigma^1, \widetilde{\sigma}^2(k)) = \{g(\sigma^1, \widetilde{\sigma}^2(k)) \mid \sigma^1 \in \Sigma^1\} = [(0,1),(\widetilde{x}(k),0)] \,. \tag{7.3}$$

Let $\widetilde{\sigma}^2(k_j)$ with $\widetilde{\sigma}^2(k_j) \to \sigma_0^2$ for $j \to \infty$ be a converging subsequence. Then

$$g(\Sigma^1, \sigma_0^2) = [(1,0),(0,1)] \,. \tag{7.4}$$

Denote $\sigma^2(j) = \widetilde{\sigma}^2(k_j)$ and $x(j) = \widetilde{x}(k_j)$, $j = 1, 2, \ldots$
By (7.2) and (7.3)

$$g\left(\sigma_0^1, \sigma^2(j)\right) = (0,1), \;\; j = 1, 2, \ldots$$

Hence, $g(\sigma_0^1, \sigma_0^2) = \lim_{j\to\infty} g\left(\sigma^1, \sigma^2(j)\right) = (0,1)$. On the other hand, by (7.1) and (7.4)

$$g\left(\sigma^1(j), \sigma_0^2\right) = (1,0), \;\; j = 1, 2, \ldots$$

Hence, $g(\sigma_0^1, \sigma_0^2) = \lim_{j\to\infty} g\left(\sigma^1(j), \sigma_0^2\right) = (1,0)$. Thus, we have a contradiction as desired.

The foregoing example indicates that the main reason for non-existence of continuous presentations is the discontinuity of set intersection. In order make this precise we introduce the Hausdorff metric $d_H$ on $\mathcal{K}(A)$. Let $d$ be a metric on $A$. For $B \in \mathcal{K}(A)$ and $\varepsilon > 0$ we denote

$$\overline{U}(B,\varepsilon) = \{x \in A \mid \text{there exist } y \in B \text{ such that } d(x,y) \leq \varepsilon\} .$$

For $B_1, B_2 \in \mathcal{K}(A)$ we define

$$d_H(B_1, B_2) = \inf\{\varepsilon > 0 \mid B_2 \subseteq \overline{U}(B_1,\varepsilon) \text{ and } B_1 \subseteq \overline{U}(B_2,\varepsilon)\} .$$

One can prove that $d_H$ is a metric, the *Hausdorff metric*, and $(\mathcal{K}(A), d_H)$ is a compact metric space (see Hildenbrand, 1974, p. 17).

Return to Example 7.2.1 and let now $\alpha(k) \uparrow 1$. Then $[(1,0),(0,\alpha(k))] \to [(1,0),(0,1)]$ and $[(\alpha(k),0),(0,1)] \to [(1,0),(0,1)]$ in $(\mathcal{K}(A), d_H)$. However,

$$[(1,0),(0,\alpha(k))] \cap [(\alpha(k),0),(0,1)] = \left(\frac{\alpha(k)}{1+\alpha(k)}, \frac{\alpha(k)}{1+\alpha(k)}\right) \to \left(\frac{1}{2}, \frac{1}{2}\right) .$$

Thus, the limit of the intersections is strictly contained in the intersection of the limits.

## 7.3 Finitely generated effectivity functions and $\varepsilon$-representations

In view of Example 7.2.1 in the preceding section we have to impose extra conditions on an effectivity function in order to obtain a continuous representation. A possible simple condition is dependence on a finite number of sets of alternatives. This is made precise in the following.

Let $A$ be a compact metric space.

**Definition 7.3.1.** An effectivity function $E : P(N) \to P(\mathcal{K}(A))$ is *finitely generated* if $E(\emptyset) = \emptyset$, $E(N) = \mathcal{K}(A)$, and for every $S \subseteq N$, $S \neq \emptyset, N$, there exist $k(S) \in \mathbb{N}$ and $B(j,S) \in \mathcal{K}(A)$, $1 \leq j \leq k(S)$, such that

$$E(S) = \{B \in \mathcal{K}(A) \mid B \supseteq B(j,S) \text{ for some } 1 \leq j \leq k(S)\} . \tag{7.5}$$

With this definition we can formulate our first existence result.

**Theorem 7.3.2.** *Let $E : P(N) \to P(\mathcal{K}(A))$ be a monotonic and superadditive effectivity function. Let $E$ be finitely generated, as in Definition 7.3.1. Then $E$ has a continuous representation.*

*Proof.* The sets $B(j,S)$, $j = 1,\ldots,k(S)$, $S \neq \emptyset, N$, generate a finite algebra $\mathcal{F}$. Denote by $\widehat{A}$ the atoms (i.e., minimal elements) of $\mathcal{F}$. For each $B \in P_0(\widehat{A})$ denote

$$\varphi(\widehat{B}) = \bigcup\{\widehat{a} \mid \widehat{a} \in \widehat{B}\} \in P_0(A) . \tag{7.6}$$

This enables us to define a discrete effectivity function $\widehat{E} : P(N) \to P(P_0(\widehat{A}))$ by

$$\widehat{E}(S) = \{\widehat{B} \subseteq \widehat{A} \mid \varphi(\widehat{B}) \in E(S)\} , \tag{7.7}$$

for $S \neq \emptyset, N$; $\widehat{E}(N) = P_0(\widehat{A})$; and $\widehat{E}(\emptyset) = \emptyset$. As the reader may verify easily, $\widehat{E}$ is monotonic and superadditive. Theorem 2.4.7 implies that $\widehat{E}$ has a (discrete) representation $\widehat{\Gamma} = (N; \widehat{\Sigma}^1, \ldots, \widehat{\Sigma}^n; \widehat{g}; \widehat{A})$. Call $f : \widehat{A} \to A$ a *choice function* if $f(\widehat{a}) \in \widehat{a}$ for all $\widehat{a} \in \widehat{A}$, and denote by $F$ the set of all choice functions. Finally, define a new game form $\Gamma = (N; \Sigma^1, \ldots, \Sigma^n; g; A)$ by

(i) $\Sigma^i = \widehat{\Sigma}^i \times F \times N$ for all $i \in N$;
(ii) $g\left((\widehat{\sigma}^1, f^1, t^1), \ldots, (\widehat{\sigma}^n, f^n, t^n)\right) = f^j\left(\widehat{g}(\widehat{\sigma}^1, \ldots, \widehat{\sigma}^n)\right)$, where $j = (t^1 + \ldots + t^n) \bmod n$.

If we endow $\widehat{\Sigma}^1, \ldots, \widehat{\Sigma}^n$ and $N$ with the discrete topology and $F = \prod_{\widehat{a} \in \widehat{A}} \widehat{a}$ with the product topology, then $g$ is continuous. It remains to show that $\Gamma$ is a representation of $E$. Clearly, $E^\Gamma(N) = \mathcal{K}(A)$. Thus, let $S \neq \emptyset, N$. First, let $B = B(j, S)$ for some $1 \leq j \leq k(S)$. Then there is $\widehat{B} \subseteq \widehat{A}$ such that $\varphi(\widehat{B}) = B \in E(S)$, hence $\widehat{B} \in \widehat{E}(S)$. Since $\widehat{\Gamma}$ represents $\widehat{E}$, this implies that $S$ has a strategy profile $\widehat{\sigma}^S \in \widehat{\Sigma}^S$ such that $\widehat{g}(\widehat{\sigma}^S, \widehat{\mu}^{N \setminus S}) \in \widehat{B}$ for all $\widehat{\mu}^{N \setminus S} \in \widehat{\Sigma}^{N \setminus S}$. Therefore, for arbitrary but fixed $f^S, t^S$ it follows that $g\left((\widehat{\sigma}^S, f^S, t^S), (\widehat{\mu}^{N \setminus S}, f^{N \setminus S}, t^{N \setminus S})\right) \in B$ for all $(\widehat{\mu}^{N \setminus S}, f^{N \setminus S}, t^{N \setminus S}) \in \Sigma^{N \setminus S}$. Thus, $B \in E^\Gamma(S)$, implying that $E^\Gamma(S) \supseteq E(S)$.

Second, let $D \in \mathcal{K}(A) \setminus E(S)$. Suppose that, for some $\widehat{B} \in \widehat{E}(S)$, we had $\varphi(\widehat{B}) \subseteq D$. Then, since $\varphi(\widehat{B}) \in E(S)$, we would have $D \in E(S)$ by monotonicity, a contradiction. Hence $\varphi(\widehat{B}) \setminus D \neq \emptyset$ for every $\widehat{B} \in \widehat{E}(S)$.

Consider any $\widehat{\sigma}^S \in \widehat{\Sigma}^S$. If $\widehat{g}(\widehat{\sigma}^S, \widehat{\mu}^{N \setminus S}) \subseteq D$ for all $\widehat{\mu}^{N \setminus S} \in \widehat{\Sigma}^{N \setminus S}$, then, since $\widehat{\Gamma}$ represents $\widehat{E}$, it follows from (7.7) that $D \in E(S)$, a contradiction. Hence, for every $\widehat{\sigma}^S \in \widehat{\Sigma}^S$ there exists $\widehat{\mu}^{N \setminus S} \in \widehat{\Sigma}^{N \setminus S}$ such that $\widehat{g}(\widehat{\sigma}^S, \widehat{\mu}^{N \setminus S}) \setminus D \neq \emptyset$. Thus, in $\Gamma$ there is a player $i \in N \setminus S$ who can choose $f^i, t^i$ such that the outcome is not in $D$. Hence, $D \notin E^\Gamma(S)$ and, thus, $E^\Gamma(S) \subseteq E(S)$. □

Theorem 7.3.2 can be used to obtain an approximate representation. The family of all finitely generated effectivity functions plays an important role since it is dense within the set of all effectivity functions in a sense to be made precise below.

We start with the definition of an $\varepsilon$-representation. Recall (see Section 7.2) that in the compact metric space $A$ with metric $d$, for $B \in \mathcal{K}(A)$ and $\varepsilon > 0$ the set $\overline{U}(B, \varepsilon)$ is defined by $\overline{U}(B, \varepsilon) = \{x \in A \mid$ there exist $y \in B$ such that $d(x, y) \leq \varepsilon\}$.

**Definition 7.3.3.** Let $A$ be a compact metric space, $E : P(N) \to P(\mathcal{K}(A))$ an effectivity function, and $\varepsilon > 0$. A game form $\Gamma = (N; \Sigma^1, \ldots, \Sigma^n; g; A)$ is an *$\varepsilon$-representation* of $E$ if

(i) $E^\Gamma(S) \subseteq E(S)$ for all $S \subseteq N$;
(ii) if $S \subseteq N$ and $B \in E(S)$ then $\overline{U}(B, \varepsilon) \in E^\Gamma(S)$.

We notice now that if $\Gamma$ is an $\varepsilon$-representation of $E$, then $E^\Gamma$ is an $\varepsilon$-approximation of $E$ in $(\mathcal{K}(A), d_H)$. Indeed, for every $S \in P_0(N)$, $S \neq N$, if $B \in E(S)$ then

$$d_H(B, E^\Gamma(S)) = \inf\{d_H(B, D) \mid D \in E^\Gamma(S)\} \le d_H(B, \overline{U}(B, \varepsilon)) \le \varepsilon\ .$$

Let $E : P(N) \to P(\mathcal{K}(A))$ be a superadditive and monotonic effectivity function. Then, as we shall prove, for every $\varepsilon > 0$ there exists a continuous $\varepsilon$-representation $\Gamma_\varepsilon$ of $E$. Thus, in particular, $E$ is approximated by the effectivity functions $E^{\Gamma_\varepsilon}$, which have continuous representations.

**Theorem 7.3.4.** *Let $E : P(N) \to P(\mathcal{K}(A))$ be a monotonic and superadditive effectivity function and let $\varepsilon > 0$. Then there exists a continuous $\varepsilon$-representation of $E$.*

*Proof.* We choose for each $S \in P_0(N)$, $S \neq N$, a finite set $D^*(S) = \{B(1, S), \ldots, B(k(S), S)\} \subseteq E(S)$ such that for every $B \in E(S)$ there exists $1 \le j \le k(S)$ with $\overline{U}(B, \varepsilon) \supseteq B(j, S)$. This is possible by a standard argument, using compactness of $(\mathcal{K}(A), d_H)$. By induction on $|S|$ we construct a system $D(S)$, $S \neq \emptyset, N$, such that

(1) $D^*(S) \subseteq D(S) \subseteq E(S)$;
(2) $D(S)$ is finite;
(3) for all $S, T$: $S \subseteq T \Rightarrow D(S) \subseteq D(T)$; and
(4) if $B_i \in D(S_i)$, $i = 1, 2$, and $S_1 \cap S_2 = \emptyset$, then $B_1 \cap B_2 \in E(S_1 \cup S_2)$.

(The system $D(S)$ for all $S \neq \emptyset, N$ is constructed by taking $D^*(S)$ and adding sets in such a way as to satisfy the monotonicity and superadditivity conditions in (3) and (4), starting with the singleton coalitions.) Let now $\widetilde{E} : P(N) \to P(\mathcal{K}(A))$ be the effectivity function which is finitely generated by the system $D(S)$, $S \neq \emptyset, N$. Then $\widetilde{E}$ is monotonic and superadditive. By Theorem 7.3.2, $\widetilde{E}$ has a continuous representation $\Gamma$. By the way the sets $D^*(S)$ are chosen it follows that $\Gamma$ is an $\varepsilon$-representation of $E$. □

## 7.4 The reduction theorem

Let $A$ be a compact metric space of alternatives and let $N = \{1, \ldots, n\}$ be the set of members of the society. Suppose that there exists another compact metric space $M$ and a continuous surjection $f : M \to A$. Then, as we shall prove, each effectivity function $E : P(N) \to P(\mathcal{K}(A))$ can be 'lifted' to the space $M$ to yield an effectivity function $\widetilde{E} : P(N) \to P(\mathcal{K}(M))$ such that with every continuous representation of $\widetilde{E}$ we can associate in a natural way a continuous representation of $E$. This result is our 'reduction theorem'. It is useful in view of the following famous result in general topology.

**Theorem 7.4.1.** *If $A$ is a compact metric space, then there exists a continuous mapping from the Cantor set $\mathcal{C}$ onto $A$.*

*Remark 7.4.2.* Recall that the Cantor set is the subset of the unit interval $[0, 1]$ constructed as follows. From $[0, 1]$ leave out the open interval $(\frac{1}{3}, \frac{2}{3})$, from the remaining set leave out the open intervals $(\frac{1}{9}, \frac{2}{9})$ and $(\frac{7}{9}, \frac{8}{9})$, etc., *ad infinitum*. The remaining set is the Cantor set $\mathcal{C}$. Alternatively, the Cantor set is the set of numbers in $[0, 1]$ of which the triadic expansion does not contain the number 1, with the understanding that, for instance, the number $\frac{1}{3}$ is written as 0.02222... in triadic expansion. Clearly, the Cantor set is a compact and metric space. Its complement is open and dense in $[0, 1]$. See Kelley (1955, pp. 165–166) for a treatment of the Cantor set and Theorem 7.4.1.

In view of Theorem 7.4.1 it is sufficient to find continuous representations when $A = \mathcal{C}$. This, indeed, simplifies the problem since $\mathcal{K}(\mathcal{C})$ has a continuous selection: for each $B \in \mathcal{K}(\mathcal{C})$ define $\varphi(B) = \max_{x \in B} x$, then $\varphi : \mathcal{K}(\mathcal{C}) \to \mathcal{C}$ is continuous (where $\mathcal{K}(\mathcal{C})$ is endowed with the Hausdorff metric). This fact will be used in the next section.

We now turn to the general reduction theorem. As above, let $A$ and $M$ be compact metric spaces and let $f : M \to A$ be a continuous surjection. Let $E : P(N) \to P(\mathcal{K}(A))$ be an effectivity function, then we define a function $\widetilde{E} : P(N) \to P(\mathcal{K}(M))$ by $\widetilde{E}(\emptyset) = \emptyset$, $\widetilde{E}(N) = \mathcal{K}(M)$, and

$$\widetilde{E}(S) = \{\widetilde{B} \in \mathcal{K}(M) \mid \widetilde{B} \supseteq f^{-1}(B) \text{ for some } B \in E(S)\} \tag{7.8}$$

otherwise. Clearly, $\widetilde{E}$ is an effectivity function.[1] Also, we note the following result.

**Lemma 7.4.3.** *If $E$ is monotonic and superadditive, then $\widetilde{E}$ is also monotonic and superadditive.*

*Proof.* Assume that $E$ is monotonic and superadditive. Then the monotonicity of $\widetilde{E}$ is obvious. We are left to prove superadditivity. Let $\widetilde{B}_i \in \widetilde{E}(S_i)$, $i = 1, 2$, and $S_1 \cap S_2 = \emptyset$. There exist $B_i \in E(S_i)$, $i = 1, 2$, such that $\widetilde{B}_i \supseteq f^{-1}(B_i)$, $i = 1, 2$. By superadditivity of $E$, $B_1 \cap B_2 \in E(S_1 \cup S_2)$. Hence

$$\widetilde{B}_1 \cap \widetilde{B}_2 \supseteq f^{-1}(B_1) \cap f^{-1}(B_2) = f^{-1}(B_1 \cap B_2)$$

and therefore $\widetilde{B}_1 \cap \widetilde{B}_2 \in \widetilde{E}(S_1 \cup S_2)$. □

We now assume that $E$ and, thus, $\widetilde{E}$ are monotonic and superadditive, and remark for future reference that

$$[\widetilde{B} \in \widetilde{E}(S) \Rightarrow f(\widetilde{B}) \in E(S)] \text{ for all } S \in P(N) \text{ and } \widetilde{B} \in \widetilde{E}(S). \tag{7.9}$$

The main result of this section is the following.

**Theorem 7.4.4 (Reduction theorem).** *Let $\widetilde{\Gamma} = (N; \Sigma^1, \ldots, \Sigma^n; g; M)$ be a (continuous) representation of $\widetilde{E}$. Then $\Gamma = (N; \Sigma^1, \ldots, \Sigma^n; f \circ g; A)$ is a (continuous) representation of $E$.*

---

[1] Not to be confused with the lottery model in Chapter 6, for which the same notation was used.

*Proof.* Since $f \circ g$ is continuous if $g$ is continuous, we only have to prove that $\Gamma$ is a representation of $E$. Let $S \in P_0(N)$ and $B \in E(S)$. Then $f^{-1}(B) \in \widetilde{E}(S)$, and therefore there exists $\sigma^S \in \Sigma^S$ such that $g(\sigma^S, \mu^{N \setminus S}) \in f^{-1}(B)$ for all $\mu^{N \setminus S} \in \Sigma^{N \setminus S}$, i.e., $f(g(\sigma^S, \mu^{N \setminus S})) \in B$. Thus, $B \in E^\Gamma(S)$.

Let now $D \in \mathcal{K}(A) \setminus E(S)$. By (7.9), $f^{-1}(D) \notin \widetilde{E}(S)$. Hence, for every $\sigma^S \in \Sigma^S$ there exists $\mu^{N \setminus S} \in \Sigma^{N \setminus S}$ such that $g(\sigma^S, \mu^{N \setminus S}) \notin f^{-1}(D)$. Thus, $f(g(\sigma^S, \mu^{N \setminus S})) \notin D$, and $D \notin E^\Gamma(S)$. □

*Remark 7.4.5.* An interesting consequence of Example 7.2.1 and Theorems 7.4.1 and 7.4.4 is the existence of an effectivity function $E : P(\{1,2\}) \to P(\mathcal{K}(\mathcal{C}))$ that has no continuous representation.

## 7.5 Semicontinuous representations on $\mathbb{R}$

In view of the reduction theorem (Theorem 7.4.4) and Theorem 7.4.1, which says that each compact metric space is a continuous image of the Cantor set, the latter set will play a special role in our investigation of the existence of continuous representations. In this section we consider the case where the set of alternatives $A$ is a compact subset of the real line. In that case it makes sense to consider upper or lower semicontinuity of outcome functions.

We start with a remark about the relation between the upper topology (see Section 3.4) and the topology induced by the Hausdorff distance (see Section 7.2) on the set of nonempty closed subsets of a compact metric space.

*Remark 7.5.1.* Let $A$ be a compact metric space with metric $d$ and let $K = \{B \in \mathcal{K}(A) \mid B \subseteq U\}$, where $U$ is an open set in $A$, be a base element of the upper topology. Clearly, $K$ is also open in $(\mathcal{K}(A), d_H)$, so that the upper topology is a coarsening of the topology induced by the Hausdorff metric: every subset of $\mathcal{K}(A)$ that is open [closed] in the upper topology is also open [closed] in the topology induced by the Hausdorff topology.

Conversely, suppose that $K \subseteq \mathcal{K}(A)$ is closed in $(\mathcal{K}(A), d_H)$, and that $K$ is closed under taking supersets, that is, $B \in K$ implies $B' \in K$ for all $B, B' \in \mathcal{K}(A)$ with $B \subseteq B'$. Then $K$ is also closed in the upper topology. To see this, let $B \in \mathcal{K}(A)$ be a limit point of $K$ in the upper topology. For every $k \in \mathbb{N}$ define the open (in $A$) set $U_k$ by $U_k = \{y \in A \mid d(y, B) < \frac{1}{k}\}$. Then the set $\{C \in \mathcal{K}(A) \mid C \subseteq U_k\}$ is an open neighborhood of $B$ in the upper topology, and therefore contains an element $B_k \in K$. Every limit point of a subsequence of $(B_k)_{k \in \mathbb{N}}$ in $(\mathcal{K}(A), d_H)$ is in the closure (within $(\mathcal{K}(A), d_H)$) of $K$, hence in $K$ since $K$ is by assumption closed in $(\mathcal{K}(A), d_H)$. Moreover, by definition of $U_k$, every such limit point is a subset of $B$. Since $K$ is closed under taking supersets, it follows that $B \in K$. Hence, $K$ is closed in the upper topology.

In particular, suppose that $E : P(N) \to P(\mathcal{K}(A))$ is monotonic and closed-valued, that is, for every $S \subseteq N$ the set $E(S)$ is closed in $(\mathcal{K}(A), d_H)$. Then $E(S)$ is closed in the upper topology for every $S \subseteq N$.

The following theorem provides sufficient conditions for the existence of a semicontinuous representation when the set of alternatives is a compact subset $A$ of the real line. We say that $\Gamma = (N; \Sigma^1, \ldots, \Sigma^n; g; A)$ is a *semicontinuous* representation if the outcome function $g : \Sigma \to A$ is upper or lower semicontinuous. (Given a topology on $\Sigma$, the function $g : \Sigma \to A$ is upper [lower] semicontinuous whenever $\sigma_t \to \sigma$ implies $\limsup_{t\to\infty} g(\sigma_t) \leq g(\sigma)$ [$\liminf_{t\to\infty} g(\sigma_t) \geq g(\sigma)$].)

**Theorem 7.5.2.** *Let $A$ be a compact subset of $\mathbb{R}$ and let $E : P(N) \to P(\mathcal{K}(A))$ be a monotonic and superadditive effectivity function, such that $E(S)$ is closed in $(\mathcal{K}(A), d_H)$ for every $S \subseteq N$. Then $E$ has a semicontinuous representation.*

*Proof.* We will construct a representing game form with upper semicontinuous outcome function. Let $N^i = \{S \subseteq N \mid i \in S\}$. Now let

$$\Upsilon^i = \{\upsilon^i : N^i \to N^i \times N \mid \upsilon^i_1(S) \subseteq S \text{ and } \upsilon^i_2(S) \in S\} ,$$

where $\upsilon^i = (\upsilon^i_1, \upsilon^i_2)$. Further, let

$$M^i = \{\psi^i : N^i \to \mathcal{K}(A) \mid \psi^i(S) \in E(S) \text{ for every } S \in N^i\} ,$$

and

$$M^i_* = \{\psi^i_* : N^i \to \mathcal{K}(A) \mid \psi^i_*(S) \in E^*(S) \text{ for every } S \in N^i\} .$$

Define now a game form $\Gamma = (N; \Sigma^1, \ldots, \Sigma^n; g; A)$ by the following rules. Let $\Sigma^i = \Upsilon^i \times M^i \times M^i_* \times N \times \{0, 1\}$ for all $i \in N$. Here, $M^i = \prod_{S^i \in N^i} E(S)$ and $M^i_* = \prod_{S^i \in N^i} E^*(S)$ are endowed with the product topology, and $\Upsilon^i$, $N$, and $\{0, 1\}$ are endowed with the discrete topology. Observe that, since $E^*(S)$ is closed in the upper topology (see Section 3.4), by Remark 7.5.1 it is also closed in $(\mathcal{K}(A), d_H)$, for each $S \subseteq N$. Thus, $\Sigma^i$ is a compact metric space for every $i \in N$.

It remains to define $g$. Let $\sigma^i = (\upsilon^i, \psi^i, \psi^i_*, t^i, q^i)$ for $i \in N$. Using $\upsilon^1, \ldots, \upsilon^n$ we introduce the following partitions of $N$. First, for $S \in P_0(N)$, we define an equivalence relation $\sim_\sigma$ on $S$ by

$$i \sim_\sigma j \Leftrightarrow \upsilon^i(S) = \upsilon^j(S) \text{ for all } i, j \in S ,$$

where $\sigma = (\sigma^1, \ldots, \sigma^n)$. Denote by $D(S, \sigma)$ the partition of $S$ with respect to $\sim_\sigma$. Let the first partition of $N$ be $H_0(\sigma) = \{N\}$ and define inductively the following partitions, as follows. If $H_k(\sigma) = \{S_{k,1}, \ldots, S_{k,\ell}\}$ is the $k$-th partition, $k \geq 0$, then we define

$$H_{k+1}(\sigma) = \bigcup_{j=1}^{\ell} D(S_{k,j}, \sigma).$$

Clearly, there exists a minimal $r$ such that $H_k(\sigma) = H_r(\sigma)$ for all $k \geq r$. Let $H_r = \{S_1, \ldots, S_\ell\}$. The coalitions $S_1, \ldots, S_\ell$ are called *final*. For each final coalition $S_j$, $j = 1, \ldots, \ell$, there exists $k_j \in S_j$ such that $v^i(S_j) = (S_j, k_j)$ for all $i \in S_j$. Further, a final coalition is called *decided* if $q^{k_j} = 0$.

In defining $g$ we distinguish the following three cases:

$$\ell = 1. \text{ Then let } g(\sigma) = \max(\psi^{k_1}(N)) \; (= \max\{x \mid x \in \psi^{k_1}(N)\}). \tag{7.10}$$

$$\begin{cases} \ell > 1 \text{ and } S_1, \ldots, S_\ell \text{ are decided.} \\ \text{Then let } g(\sigma) = \max\left(\psi^{k_1}(S_1) \cap \ldots \cap \psi^{k_\ell}(S_\ell)\right). \end{cases} \tag{7.11}$$

$$\begin{cases} S_1, \ldots, S_h,\ h \geq 1, \text{ are undecided and } S_{h+1}, \ldots, S_\ell \text{ are decided.} \\ \text{Then choose } j = \sum_{u=1}^{h} t^{k_u} \bmod h, \text{ and let} \\ \quad g(\sigma) = \max\left(\bigcap_{u \neq j} \psi^{k_u}(S_u) \cap \psi_*^{k_j}(S_j)\right). \end{cases} \tag{7.12}$$

We claim that $g$ is upper semicontinuous. To this end we first observe the following inclusion.

$$B_i(t) \to B_i \text{ in } (\mathcal{K}(A), d_H),\ i = 1, 2 \Rightarrow \limsup_{t \to \infty} B_1(t) \cap B_2(t) \subseteq B_1 \cap B_2\,. \tag{7.13}$$

(If $B(t) \in \mathcal{K}(A)$, $t = 1, 2, \ldots$ then $x \in \limsup_{t \to \infty} B(t)$ if there exists a subsequence $t_1 < t_2 < \ldots$ and $x(t_j) \in B(t_j)$, $j = 1, 2, \ldots$, such that $x = \lim_{j \to \infty} x(t_j)$.)

Let now $\sigma_m^i = (v_m^i, \psi_m^i, \psi_{*m}^i, t_m^i, q_m^i) \to \sigma^i$ for $m \to \infty$, $i \in N$. Then there is $m_0$ such that $v_m^i, t_m^i, q_m^i$, $i = 1, \ldots, n$, are constant for $m \geq m_0$. This implies that for $m \geq m_0$, the determination of $g$ stays in the same case (7.10), (7.11), or (7.12); neither the attained partition nor decidedness or undecidedness change; and the players making the choices do not change. Only the sets of alternatives from which the choice is made change. Now the upper semicontinuity of $g$ follows with the aid of (7.13).

We shall now prove that $\Gamma$ is a representation of $E$. Let $S \in P_0(N)$, $S \neq N$, let $B \in E(S)$ and let $u \in S$. Let $\sigma^S \in \Sigma^S$ satisfy $v^i(S') = (S, u)$ and $\psi^i(S') = B$ for all $i \in S$ and $S' \supseteq S$, and, in addition, $q^u = 0$. Then for every $\mu^{N \setminus S} \in \Sigma^{N \setminus S}$, $S$ is a decided final coalition, and $\psi^u(S) = B$. Hence, $g(\sigma^S, \mu^{N \setminus S}) \in B$ for every $\mu^{N \setminus S} \in \Sigma^{N \setminus S}$. Thus, $E(S) \subseteq E^\Gamma(S)$.

Next, let $D \in \mathcal{K}(A) \setminus E(S)$. By Remark 7.5.1 $E(S)$ is closed in the upper topology for every $S \subseteq N$. Hence, by Lemma 4.2.8, $E = E^{**}$.[2] Thus, $D \notin E^{**}(S)$. Therefore, by definition of the polar, there exists $B \in E^*(N \setminus S)$ such that $B \cap D = \emptyset$. Let $u \in N \setminus S$, let $\mu^{N \setminus S} \in \Sigma^{N \setminus S}$ satisfy $v^i(T) = (N \setminus S, u)$ and $\psi_*^i(T) = B$ for all $i \in N \setminus S$ and $T \supseteq N \setminus S$. Further, let $q^u = 1$. For any $\sigma^S \in \Sigma^S$, $N \setminus S$ is a final undecided coalition with respect to $(\sigma^S, \mu^{N \setminus S})$, and $\psi_*^u(N \setminus S) = B$. By adjusting $t^u$ (after $\sigma^S$ is chosen), $N \setminus S$ can arrange that $g(\sigma^S, \mu^{N \setminus S}) \in B$. Since $B \cap D = \emptyset$, this implies $D \notin E^\Gamma(S)$. Thus, $E^\Gamma(S) \subseteq E(S)$. □

---

[2] That is, $E$ is *reflexive*.

The fact that the outcome function $g$ in the game form $\Gamma$ constructed in the proof of Theorem 7.5.2 is only semicontinuous but does not have to be continuous, is due to the lack of continuity of set intersection, cf. Example 7.2.1. Although in this example the set of alternatives is a subset of $\mathbb{R}^2$, it is not difficult to give an example where $A \subseteq \mathbb{R}$, see Remark 7.4.5 – where $A$ is the Cantor set – or the following example.

*Example 7.5.3.* Let $N = \{1, 2\}$, $A = [0, 2]$, and let $E : P(N) \to P(\mathcal{K}(A))$ be defined by $E(\emptyset) = \emptyset$; $E(N) = \mathcal{K}(A)$; $E(1)$ contains all sets of the form

$$B_1^k = \left\{\frac{1}{k}, \frac{2}{k}, \frac{3}{k}, \ldots, \frac{2k-1}{k}\right\},\ k = 1, 2, \ldots$$

and all closed supersets within $[0, 2]$; and $E(2)$ contains all sets of the form

$$B_2^k = \left\{1, \frac{1}{k\sqrt{2}}, \frac{2}{k\sqrt{2}}, \frac{3}{k\sqrt{2}}, \ldots, \frac{m_k}{k\sqrt{2}}\right\},\ \frac{m_k}{k\sqrt{2}} < 2 < \frac{m_k + 1}{k\sqrt{2}},\ k = 1, 2, \ldots$$

and all closed supersets within $[0, 2]$. Then $E$ is a superadditive and monotonic effectivity function. Moreover, $B_1^k \cap B_2^k = \{1\}$ for each $k = 1, 2, \ldots$, whereas $\lim_{k\to\infty} B_1^k = \lim_{k\to\infty} B_2^k = [0, 2]$. Hence $\lim_{k\to\infty} B_1^k \cap B_2^k = \{1\} \neq [0, 2] = \lim_{k\to\infty} B_1^k \cap \lim_{k\to\infty} B_2^k$. (Limits are taken with respect to the Hausdorff metric.)

Suppose that in the game form $\Gamma$ constructed in the proof of Theorem 7.5.2 the two players play strategies $\sigma_k^1$ and $\sigma_k^2$, $k = 1, 2, \ldots$, such that $g(\sigma_k) = g(\sigma_k^1, \sigma_k^2) = \max\left(B_1^k \cap B_2^k\right)$, i.e., case (7.11) applies for each $k = 1, 2, \ldots$ Then $\lim_{k\to\infty} g(\sigma_k) = \lim_{k\to\infty} \max\left(B_1^k \cap B_2^k\right) = 1$, whereas $g(\sigma) = g(\lim_{k\to\infty} \sigma_k) = \max([0, 2]) = 2$. Clearly, $g$ is not continuous.

Next we investigate Nash consistency of the game form $\Gamma$ constructed in the proof of Theorem 7.5.2. Assume that the conditions of this theorem are satisfied and denote

$$Q^N = \{R^N \in V^N \mid \exists x \in A\, \forall i \in N\, [L(x, R^i) \in E(N \setminus i)]\}\,. \tag{7.14}$$

Proposition 3.2.1 implies that $Q^N$ is the maximal subdomain of the domain of continuous preference profiles $V^N$ on which $\Gamma$ may be Nash consistent. We now prove that $\Gamma$ is indeed Nash consistent on $Q^N$.

**Theorem 7.5.4.** *Let $\Gamma$ be the game form constructed in the proof of Theorem 7.5.2. Then for every $R^N \in Q^N$, defined by (7.14), the game $(\Gamma, R^N)$ has a Nash equilibrium.*

*Proof.* Let $R^N \in Q^N$ with $x$ as in (7.14). Consider strategies $\sigma^1, \ldots, \sigma^n$ with

$$v^i(N) = (N, 1);\ \psi^i(N) = \{x\},\ \text{for all } i \in N, \tag{7.15}$$

and

$$\begin{cases} v^i(N \setminus j) = (N \setminus j, k_j) \text{ where } k_j = (j+1) \bmod n\,, \\ \psi^i(N \setminus j) = L(x, R^j),\ \text{all } i \in N \setminus j,\ j = 1, \ldots, n. \end{cases} \tag{7.16}$$

Then $g(\sigma) = x$ and $g(\sigma^{N\setminus i}, \tau^i) \in L(x, R^i)$ for every $\tau^i \in \Sigma^i$ and $i \in N$. Thus, $\sigma$ is a Nash equilibrium in $(\Gamma, R^N)$. □

By similar arguments as in Chapter 3, $\Gamma$ in Theorem 7.5.2 is a 'canonical' representation, i.e., a representation with a maximum set of Nash equilibrium outcomes. Also, and as in Chapter 3, since $A$ is compact it follows that $\Gamma$ is weakly acceptable.

## 7.6 Representations of effectivity functions and modified Baire functions

Properties of representations of effectivity functions like upper or lower semicontinuity may be useful in many contexts. In this section we continue this line of investigation of continuity properties and we study the implication of Theorem 7.5.2 for general compact metric spaces, relying on a generalization of the reduction theorem (Lemma 7.6.3). We start with the following definitions.

Let $M$ be a metric space. A set $B \subseteq M$ is called a $G_\delta$ $[F_\sigma]$ if it can be written as a countable intersection of open sets [a countable union of closed sets][3]. It is a $G_{\delta\sigma}$ if it can be written as a countable union of sets, each of which is a $G_\delta$. A function $f : M \to O$ between metric spaces $M$ and $O$ is a Baire function of class 0 if it is continuous, that is, $f^{-1}(G)$ is open for every open set $G$. It is a (modified) Baire function of class 1 if $f^{-1}(G)$ is an $F_\sigma$ for each open set $G$, and, finally, it is a (modified) Baire function of class 2 if $f^{-1}(G)$ is a $G_{\delta\sigma}$ for each open set $G$. For further discussion of Baire classes, the reader is referred to Hausdorff (1962).

Let $M$ be a compact metric space and let $E : P(N) \to P(\mathcal{K}(M))$ be a monotonic and superadditive effectivity function with closed values (in $(\mathcal{K}(M), d_H)$). By Lemma 7.4.3, $E$ can be 'lifted' to a monotonic and superadditive effectivity function $\widetilde{E} : P(N) \to P(\mathcal{K}(\mathcal{C}))$ as defined in (7.8). We will prove below that $\widetilde{E}$ may be chosen to have closed values in $(\mathcal{K}(\mathcal{C}), d_H)$. Therefore, by Theorem 7.5.2, $\widetilde{E}$ has an upper semicontinuous representation $\widetilde{\Gamma} = (N; \Sigma^1, \ldots, \Sigma^n; g; \mathcal{C})$. Applying now the reduction theorem (Theorem 7.4.4) we obtain that $\Gamma = (N; \Sigma^1, \ldots, \Sigma^n; f \circ g; M)$ is a representation of $E$, where $f : \mathcal{C} \to M$ is a continuous surjection of the Cantor set onto $M$. Now, if $U \subseteq M$ is an open set, then $(f \circ g)^{-1}(U) = g^{-1}(f^{-1}(U))$ is, as we shall prove, a $G_{\delta\sigma}$ of $\Sigma = \Sigma^1 \times \ldots \times \Sigma^n$. Hence, $f \circ g$ is a modified Baire function of class 2 – see Appendix D in Hausdorff (1962).

We summarize the foregoing discussion in the following theorem.

**Theorem 7.6.1.** *Let $M$ be a compact metric space and let $E : P(N) \to P(\mathcal{K}(M))$ be a monotonic and superadditive effectivity function with closed values in $(\mathcal{K}(M), d_H)$. Then $E$ has a representation $\Gamma = (N; \Sigma^1, \ldots, \Sigma^n; \widehat{g}; M)$ such that for every open set $U \subseteq M$ we have that $\widehat{g}^{-1}(U)$ is a $G_{\delta\sigma}$, that is, $\widehat{g}$ is a modified Baire function of class 2.*

---

[3] $G_\delta$ sets were discussed before, see Section 5.4.

The proof of Theorem 7.6.1 follows from the lemmas below. Let $M$ and $E$ as in the theorem and let $f : \mathcal{C} \to M$ be a continuous surjection. Define an effectivity function $\widetilde{E} : P(N) \to P(\mathcal{K}(\mathcal{C}))$ in two steps. For $S \in P_0(N)$, $S \neq N$,

$$\widehat{E}(S) = \mathrm{cl}\,\{\widehat{B} \in \mathcal{K}(\mathcal{C}) \mid \widehat{B} = f^{-1}(B) \text{ for some } B \in E(S)\}\ ; \tag{7.17}$$

$$\widetilde{E}(S) = \{\widetilde{B} \in \mathcal{K}(\mathcal{C}) \mid \widetilde{B} \supseteq \widehat{B} \text{ for some } \widehat{B} \in \widehat{E}(S)\}\ . \tag{7.18}$$

As usual, $\widetilde{E}(N) = \mathcal{K}(\mathcal{C})$ and $\widetilde{E}(\emptyset) = \emptyset$.

**Lemma 7.6.2.** *$\widetilde{E}$ is a monotonic and superadditive effectivity function with closed values.*

*Proof.* Monotonicity is obvious. Also, $\widetilde{E}$ has closed values since $\widehat{E}$ has (by definition) closed values. For superadditivity, let $S_1, S_2 \in P_0(N)$, $S_1 \cap S_2 = \emptyset$, $\widehat{B}_1 \in \widehat{E}(S_1)$, and $\widehat{B}_2 \in \widehat{E}(S_2)$. Then there exist $B_{1,k} \in E(S_1)$, $B_{2,k} \in E(S_2)$, $k = 1, 2, \ldots$ such that $f^{-1}(B_{1,k}) \to \widehat{B}_1$ and $f^{-1}(B_{2,k}) \to \widehat{B}_2$. Clearly, $C_k = B_{1,k} \cap B_{2,k} \in E(S_1) \cup E(S_2)$, $k = 1, 2 \ldots$ We may assume (by considering a subsequence) that $f^{-1}(C_k) \to \widehat{C}$. By (7.17), $\widehat{C} \in \widehat{E}(S_1 \cup S_2)$. Also, $f^{-1}(C_k) = f^{-1}(B_{1,k}) \cap f^{-1}(B_{2,k})$ for each $k$. Hence, $\widehat{C} \subseteq \widehat{B}_1 \cap \widehat{B}_2$ and, thus, $\widehat{B}_1 \cap \widehat{B}_2 \in \widetilde{E}(S_1 \cup S_2)$. □

We now observe that

$$\widetilde{B} \in \widetilde{E}(S) \Rightarrow f(\widetilde{B}) \in E(S), \text{ for all } S \in P_0(N) \text{ and } \widetilde{B} \in \mathcal{K}(\mathcal{C}). \tag{7.19}$$

Indeed, if $\widehat{B} \in \widehat{E}(S)$, then there is a sequence $(B_k)$ in $E(S)$ such that $f^{-1}(B_k) \to \widehat{B}$. By the (uniform) continuity of $f$, it follows that $B_k = f(f^{-1}(B_k)) \to f(\widehat{B})$. Thus, $f(\widehat{B}) \in E(S)$. Hence, (7.19) holds as well.

Lemma 7.6.2 and (7.19) imply the following result by an argument analogous to the proof of the reduction theorem (Theorem 7.4.4).

**Lemma 7.6.3.** *If $\widetilde{\Gamma} = (N; \Sigma^1, \ldots, \Sigma^n; g; \mathcal{C})$ is a representation of $\widetilde{E}$, then $\Gamma = (N; \Sigma^1, \ldots, \Sigma^n; f \circ g; M)$ is a representation of $E$.*

Our final lemma is a standard exercise on semicontinuous real functions.

**Lemma 7.6.4.** *Let $M^*$ be a metric space and let $g : M^* \to \mathcal{C}$ be an upper semicontinuous function. Then for every open set $U \subseteq \mathcal{C}$, $g^{-1}(U)$ is a $G_{\delta\sigma}$.*

*Proof.* First we compute the inverse image of a 'ray' in $\mathcal{C}$. There are four possibilities.

(1) $D = \{x \in \mathcal{C} \mid x \geq a\}$ for some $a \in \mathcal{C}$. Then upper semicontinuity of $g$ implies that $g^{-1}(D)$ is closed.

(2) $D = \{x \in \mathcal{C} \mid x > a\}$ for some $a \in \mathcal{C}$, $a < 1$, and $D$ is not of the form as in (1). Then there are countable $a_s \in \mathcal{C}$ such that $a_s \downarrow a$. Hence

$$g^{-1}(D) = g^{-1}\left(\bigcup_s \{x \in \mathcal{C} \mid x \geq a_s\}\right) = \bigcup_s g^{-1}\left(\{x \in \mathcal{C} \mid x \geq a_s\}\right)$$

which is an $F_\sigma$.
(3) $D = \{x \in \mathcal{C} \mid x < a\}$ for some $a \in \mathcal{C}$. Then by (1), $g^{-1}(D)$ is open.
(4) $D = \{x \in \mathcal{C} \mid x \leq a\}$ for some $a \in \mathcal{C}$. Then, for the complement $M^* \setminus g^{-1}(D)$ of $g^{-1}(D)$, either case (1) applies, so that $g^{-1}(D)$ is open; or case (2) applies, so that $g^{-1}(D)$ is a $G_\delta$ (since $M^* \setminus g^{-1}(D)$ is an $F_\sigma$).

By (1)–(4), the inverse image under $g$ of every interval (closed, open, or half-closed) in $\mathcal{C}$ is the intersection of an $F_\sigma$ with a $G_\delta$, and hence it is a $G_{\delta\sigma}$. Formally, every open set in $\mathcal{C}$ is the union of countably many intervals. Hence, if $U \subseteq \mathcal{C}$ is open, then $g^{-1}(U)$ is a $G_{\delta\sigma}$. □

Theorem 7.6.1 now follows from Lemmas 7.6.2, 7.6.3, 7.6.4, and Theorem 7.5.2.

## 7.7 Notes and comments

This chapter is based on Keiding and Peleg (2006b). It is worthwhile to emphasize that in spite of Examples 7.2.1 and 7.5.3 continuity properties of effectivity functions are latent everywhere in our model. Indeed, we prove existence of $\varepsilon$-representations of monotonic and superadditive toplological effectivity functions (Theorem 7.4.4). Also, if the space of alternatives is the Cantor set, then every monotonic and superadditive effectivity function with closed values has a semicontinuous representation (Theorem 7.5.2).

*Remark 7.7.1.* We reemphasize that if set intersection were continuous (in $(\mathcal{K}(A), d_H)$), then one could prove existence of continuous representations in $\mathcal{C}$ – this is apparent from the proof of Theorem 7.5.2. By the reduction theorem all (monotonic and superadditive) effectivity functions with closed values, on arbitrary compact metric spaces of alternatives, would have continuous representations. Thus, indeed, the discontinuity of the intersection is the sole reason for the nonexistence of continuous representations.

# Part II
# Consistent voting

# Chapter 8
# Introduction to Part II

## 8.1 Motivation and summary

In this chapter we first recall the Gibbard-Satterthwaite Theorem and review some of its implications. This is done in Section 8.2. The rest of the chapter is devoted to consideration of the problem of preference distortion as a consequence of manipulation of non-dictatorial voting rules. First we observe that the Gibbard-Satterthwaite Theorem does not tell us whether or not the sincere outcome is obtained after manipulation. It may be the case that strategic voting leads to an equilibrium of which the outcome is the sincere outcome. In that case, the result of voting by a secret ballot would be indistinguishable from that of sincere voting. Indeed, in Section 8.3 we define exactly and strongly consistent social choice functions. Such social choice functions have for each profile of (true) preferences a strong (Nash) equilibrium that yields the sincere outcome. This class of social choice functions is the main topic of Chapters 9–11, in which we present several existence and characterization theorems. In particular, in Chapter 11 we extend some of the results of Chapters 9 and 10 to voting games with a continuum of voters.

In Section 8.4 we very briefly discuss voting on restricted domains for which the manipulation problem is eliminated, and mention a few references in order to direct the reader to some of the main results in this area.

We conclude the chapter with a discussion of equilibrium with threats, following an idea of Pattanaik (1976), which presents another way to obtain the sincere outcome if manipulation is possible. We construct non-dictatorial social choice functions with the property that sincere voting is always an equilibrium with threats.

B. Peleg and H. Peters, *Strategic Social Choice*, Studies in Choice and Welfare, DOI 10.1007/978-3-642-13875-1_8, 

## 8.2 The Gibbard-Satterthwaite Theorem and its implications

Let $A$ be a set of $m$ alternatives, $m \geq 3$, and let $N = \{1, \ldots, n\}$ be a set of voters. Let $L$ denote the set of all linear orderings (strict preferences) on $A$, that is, the set of all transitive, reflexive, antisymmetric and complete binary relations on $A$. A *social choice function* (SCF) is a map $F : L^N \to A$. An SCF $F$ is *non-manipulable* (or *strategy-proof*) if for each profile of preferences $R^N \in L^N$ the strategy-profile $R^N$ itself is a Nash equilibrium in the game $(F, R^N)$. Thus, if $F$ is *manipulable* (not non-manipulable) then there exist $R_0^N \in L^N$, $i \in N$, and $Q^i \in L$ such that $F(R_0^{N \setminus \{i\}}, Q^i)\, R_0^i\, F(R_0^N)$, and $F(R_0^{N \setminus \{i\}}, Q^i) \neq F(R_0^N)$. In this case, $R_0^N$ is a situation in which player $i$ has an incentive to misrepresent his preference – 'play' $Q^i$ instead of his true preference $R_0^i$. In slightly different words, non-manipulability of a social choice function means that for every voter (player) reporting (playing) his true preference is a weakly dominant strategy in every situation, i.e., every game $(F, R^N)$.

For a social choice function $F : L^N \to A$ let the *range* of $F$, $A^*$, be defined by

$$A^* = \{x \in A \mid x = F(R^N) \text{ for some } R^N \in L^N\}.$$

A player $d \in N$ is a *dictator* of $F$ if $F(R^N)\, R^d\, x$ for every $R^N \in L^N$ and $x \in A^*$. The SCF $F$ is *dictatorial* if it has a dictator. A fundamental result of Gibbard (1973) and Satterthwaite (1975) is the following theorem.[1]

**Theorem 8.2.1.** *If a social choice function $F$ is non-manipulable and $|A^*| \geq 3$, then $F$ is dictatorial.*

Thus, if a non-dictatorial social choice function $F$ has full range ($A = A^*$) then it must be manipulable. This implies that most social choice functions based on voting procedures in every-day use, like choice by plurality voting, Borda count, and approval voting, are manipulable. A natural question is to which extent manipulability of a social choice function is a drawback. After all, a voting game $(F, R^N)$ is just a strategic game like many other every-day games (auctions, oligopolies, etc.) and so 'strategic' behavior is 'all in the game'. Nevertheless, manipulating behavior in strategic voting situations has some disturbing consequences, as pointed out by many authors (e.g., recently, Feldman and Serrano, 2005). Here, we list what we think are its main drawbacks.

First, a specific social choice function may have been adopted because of certain appealing properties, but these may be lost due to manipulation. For example, a social choice function based on plurality voting is Paretian. However, games associated with it may have equilibria resulting in outcomes that are not Pareto optimal. Cf. Feldman and Serrano (2005).

---

1 For a proof see Chap. 11 in Peters (2008).

Second, manipulation may be objectionable on ethical grounds. Specifically, a manipulating voter may benefit at the expense of others who do vote truthfully.

Third, by manipulating behavior, especially of large groups of voters, the outcome of the voting procedure may be very far from the *sincere outcome*, i.e., the outcome corresponding to the profile of true preferences. Typically, for instance, in a Parliamentary democracy, voters may not vote for a small but favored political party if that party is unlikely to be a member of the government that is formed on the basis of the national election. (More formally, assuming that voters play an equilibrium in some equilibrium correspondence $EQ$, the actual voting correspondence changes from $F$ to $F \circ EQ$.) On the basis of this argument, Feldman and Serrano (2005) question the legitimacy of the voting outcome if the voting procedure is manipulable.

Fourth, an important consequence of non-manipulability is that it requires only each voter's knowledge of his own preference. Thus, it makes the act of voting simple and reliable. This feature is lost under manipulability. On the other hand, this argument can also be considered to render the possibility of manipulation less harmful, since the cost of manipulation – e.g., to acquire the necessary information about the preferences and voting behavior of others – may prevent voters from actually manipulating.

In the next section we propose a weakening of the non-manipulability condition that takes away many of these drawbacks, since it results in the sincere outcome.

## 8.3 Exactly and strongly consistent social choice functions

In order to avoid distortion of the voting outcome implied by the Gibbard-Satterthwaite Theorem we shall weaken the non-manipulability requirement in the remainder of this book in such a way that (i) nevertheless the sincere outcome can result and (ii) this outcome can result under a strong stability condition. More precisely, we impose that the sincere outcome is always a strong (Nash) equilibrium outcome of the voting game under consideration. In a strong equilibrium (formally introduced in Definition 5.2.1) not only single players but also coalitions cannot gain by (joint) deviations. This route was first suggested in Peleg (1978a).

**Definition 8.3.1.** The social choice function $F : L^N \to A$ is *exactly and strongly consistent* (ESC) if for every $R^N \in L^N$ there exists a strong equilibrium $Q^N$ of the game $(F, R^N)$ such that $F(Q^N) = F(R^N)$.

An ESC social choice function trivially exists – take a constant social choice function, assigning a fixed alternative to any preference profile. In interesting cases, however, social choice functions are surjective, so their range is $A$. If

$F$ is a surjective ESC social choice function then, in particular, the game form $F$ is a strongly consistent representation of the effectivity function $E^F$ associated with (the game form) $F$ (see Chapter 5). This implies, in turn, that $F(Q^N)$ is an element of the core $C(E^F, R^N)$ for every strong equilibrium $Q^N$ of the game $(F, R^N)$ – see Proposition 5.2.6. It follows that the sincere outcome $F(R^N)$ of the voting game $(F, R^N)$ is in the core of $(E^F, R^N)$ and, in particular, Pareto optimal. More generally, Pareto optimality is maintained if we assume that voters play a strong equilibrium.

Moreover, under exact and strong consistency and assuming that voters play a strong equilibrium it is at least possible that the sincere outcome results. If voters have a more or less accurate conjecture about what the sincere outcome is, then a strong equilibrium $Q^N$ resulting in the sincere outcome may become a focal point in the sense of Schelling (1960) – perhaps because of ethical considerations. This may alleviate the objection of political illegitimacy of the voting procedure as mentioned in the preceding section.

As a final note, we mention that if $F$ is a surjective non-dictatorial ESC social choice function, then there exist an $R^N \in L^N$ and a strong equilibrium $Q^N$ of the game $(F, R^N)$ such that $F(Q^N) \neq F(R^N)$. Indeed, if not, then $F$ as a game form *implements* the SCF $F$ in strong equilibrium: all strong equilibria of $(F, R^N)$ result in the outcome $F(R^N)$. This implies in particular that $F$ is Maskin monotonic (see Remark 3.7.3 for the definition of Maskin monotonicity; the implication follows, e.g., by Lemma 6.5.1 in Peleg, 1984). This, in turn, implies that $F$ is dictatorial by the Muller-Satterthwaite Theorem (see Muller and Satterthwaite, 1977). In other words, under non-dictatorship we cannot have that every strong equilibrium always results in the sincere outcome.

## 8.4 Strategyproofness and restricted preferences

Following the works of Gibbard (1973) and Satterthwaite (1975) there is a large strand of literature trying to avoid the consequences of the Gibbard-Satterthwaite Theorem. Most of this literature concentrates on what is often termed the *universal domain assumption* in this theorem, which means that all (strict) preferences – both as true preferences and as reported preferences – are allowed. Dropping this assumption is usually referred to as the *restricted domain* approach.

For a social choice function $F$ one may consider the set

$$\mathrm{Sp}(F) = \{R^N \in L^N \mid R^N \text{ is a Nash equilibrium of } (F, R^N)\},$$

and then $F$ is strategy-proof on the domain $\mathrm{Sp}(F)$. The implicit assumption here is that the true preference profiles are from this domain, and that may be a strong assumption. Moreover, it may be quite difficult to compute or characterize the domain $\mathrm{Sp}(F)$.

Most results in this area are less ambitious. For instance, it was already known from Black (1948), Arrow (1951, 1963), or Dummet and Farquharson (1961) that generalized majority rule is strategy-proof when both the true and the reported preference profiles are restricted to be single-peaked. (A profile is single-peaked if there exists an ordering of the alternatives along which each individual preference is unimodal.) Blin and Satterthwaite (1976) show that this result no longer holds if the reported preference profiles are allowed to be more general, even if the true preferences are single-peaked. They do this by considering a social choice function that picks the Condorcet winner (i.e., an alternative that beats all other alternatives in pairwise comparison) if there is one, and otherwise picks the alternative with maximal Borda count.[2]

A well-known example of the converse phenomenon is approval voting, which – under some conditions on extensions of preferences from alternatives to sets of alternatives – is strategy-proof when both reported and true preferences are dichotomous: that is, each individual approves of a set $B$ and disproves of the complement (see Brams and Fishburn, 1983). In this case, strategy-proofness is lost if the true preferences can be more refined (see Roy, Peters, and Storcken, 2009).

Most results on restricted domains therefore assume that one and the same restriction applies to both the true and the reported preferences. For instance, Moulin (1980) characterized all Paretian, anonymous and strategy-proof social choice functions for a class of single-peaked preferences on the real line. For an introduction to strategy-proof social choice functions see Barberà (2001).

A different approach to the implication of the Gibbard-Satterthwaite Theorem was initiated by Kelly (1988). Accepting this implication, one may look for social choice functions that are in some way minimally manipulable, e.g., in terms of numbers of manipulable profiles[3]. See Maus, Peters, and Storcken (2007) for a recent overview.

## 8.5 Equilibrium with threats

In this section, following an idea of Pattanaik (1976), we define equilibrium with threats in game forms. We then show that there exist non-dictatorial social choice functions with the property that sincere voting is always an

[2] The Borda rule attaches a score of $m$ points to an individual's best alternative, $m-1$ points to the second best, ..., and 1 point to the worst alternative. The Borda count is obtained by summing over all individuals. At this introductory level we ignore the possibility of multiple winners.

[3] That is, social choice functions for which $\mathrm{Sp}(F)$ has maximal cardinality.

equilibrium with threats. This provides another way to cope with the negative implication of the Gibbard-Satterthwaite Theorem.

As before let $N = \{1, \ldots, n\}$ be a set of voters and let $A$ be a set of $m$ alternatives, where $n \geq 2$ and $m \geq 3$. Let $\Gamma = (\Sigma^1, \ldots, \Sigma^n; g; A)$ be a game form with surjective outcome function, let $R^N \in L^N$, and let $\sigma \in \Sigma = \prod_{i \in N} \Sigma^i$. A *threat* of a coalition $S \in P_0(N)$ against $\sigma$ is a strategy profile $\mu^S \in \Sigma^S$ such that

$$g(\mu^S, \sigma^{N \setminus S})\, R^i\, g(\sigma) \text{ for all } i \in S, \text{ and } g(\mu^S, \sigma^{N \setminus S}) \neq g(\sigma).$$

A *counter-threat* to $\mu^S$ is a strategy-profile $\mu^{N \setminus S} \in \Sigma^{N \setminus S}$ such that $g(\sigma)\, R^i\, g(\mu^S, \mu^{N \setminus S})$ for some $i \in S$. The profile $\sigma$ is an *equilibrium with threats* in $(\Gamma, R^N)$ if to each threat against $\sigma$ there exists a counter-threat. In such a strategy profile, for each deviation of a coalition $S$ that makes all its members strictly better off, there is a deviation by the complement of $S$ such that at least one member of $S$ again prefers the original outcome.

We start our discussion of equilibria with threats with the following observation.

**Lemma 8.5.1.** *Let $\Gamma = (\Sigma^1, \ldots, \Sigma^n; g; A)$ be a game form with surjective outcome function, let $R^N \in L^N$, and let $\sigma \in \Sigma$. Then $\sigma$ is an equilibrium with threats in $(\Gamma, R^N)$ if and only if $g(\sigma) \in C(E^\Gamma, R^N)$.*

*Proof.* (i) Suppose that $\sigma$ is an equilibrium with threats. If $g(\sigma) \notin C(E^\Gamma, R^N)$, then there exists $S \in P_0(N)$ and $B \in E^\Gamma(S)$ such that $B\, R^S\, g(\sigma)$ and $g(\sigma) \notin B$. As $B \in E^\Gamma(S)$ there exists $\mu_0^S \in \Sigma^S$ such that for all $\mu^{N \setminus S} \in \Sigma^{N \setminus S}$ we have $g(\mu_0^S, \mu^{N \setminus S}) \in B$. Thus, $\mu_0^S$ is a threat against $\sigma$ to which there is no counter-threat, a contradiction.

(ii) Assume that $g(\sigma) \in C(E^\Gamma, R^N)$. If $\sigma$ is not an equilibrium with threats, then there exists $S \in P_0(N)$ and $\mu_0^S \in \Sigma^S$ such that $g(\mu_0^S, \mu^{N \setminus S})\, R^i\, g(\sigma)$ and $g(\mu_0^S, \mu^{N \setminus S}) \neq g(\sigma)$ for all $i \in S$ and $\mu^{N \setminus S} \in \Sigma^{N \setminus S}$. Let $B = \{g(\mu_0^S, \mu^{N \setminus S}) \mid \mu^{N \setminus S} \in \Sigma^{N \setminus S}\}$. Then $B \in E^\Gamma(S)$ and $B\, R^S\, g(\sigma)$ with $g(\sigma) \notin B$, a contradiction. □

We now prove the claim that we made at the beginning of this section. We define a *social choice correspondence* (SCC) as a map $H : L^N \to P_0(A)$. An SCF $F$ can be seen as a special case of an SCC by setting $H_F(R^N) = \{F(R^N)\}$ for each $R^N \in L^N$. Call an SCC $H$ *citizen sovereign* if for each $a \in A$ there is an $R^N \in L^N$ with $H(R^N) = \{a\}$. With a citizen sovereign SCC $H$ we can associate an effectivity function $E^H$ by $E^H(\emptyset) = \emptyset$ and for all $S \in P_0(N)$ and $B \in P_0(A)$,

$B \in E^H(S) \Leftrightarrow$
there is $R^S \in L^S$ with $H(R^S, Q^{N \setminus S}) \subseteq B$ for all $Q^{N \setminus S} \in L^{N \setminus S}$.

Clearly, $E^H$ is superadditive.

**Theorem 8.5.2.** *There exists a nondictatorial social choice function $F$ with the property that sincere voting is always an equilibrium with threats, i.e., $R^N$ is an equilibrium with threats in the game $(F, R^N)$ for all $R^N \in L^N$.*

*Proof.* Let $E$ be a(n arbitrary) non-dictatorial, maximal and stable effectivity function. Define $F : L^N \to A$ by choosing $F(R^N) \in C(E, R^N)$ for every $R^N \in L^N$.

Observe that if $B \in P_0(A)$ and $S \in P_0(N)$ with $B \in E(S)$, then for $R^S \in L^S$ with $B\,R^S\,A \setminus B$ we have $C(E, (R^S, Q^{N\setminus S})) \subseteq B$ for all $Q^{N\setminus S} \in L^{N\setminus S}$, so that $B \in E^{C(E,\cdot)}(S)$, where $E^{C(E,\cdot)}$ is the effectivity function associated with the (citizen sovereign) social choice correspondence $C(E, \cdot)$. This implies, in turn, that $B \in E^F(S)$ (see Remark 9.3.3 in the next chapter). Hence, $E(S) \subseteq E^F(S)$ for all $S \in P_0(N)$, and since $E$ is maximal and $E^F$ superadditive, this implies $E = E^F$ by Lemma 9.3.1 in the next chapter.

Thus, $F(R^N) \in C(E^F, R^N)$ for every $R^N \in L^N$, and the proof is complete by Lemma 8.5.1. □

We conclude this section by observing that for an exactly and strongly consistent social choice function $F$ the true preference profile $R^N$ is always an equilibrium with threats of the game $(F, R^N)$. This follows from the remarks in Section 8.3, in particular from the observation that $F(R^N) \in C(E^F, R^N)$, and Lemma 8.5.1.

## 8.6 Notes and comments

The discussion in this chapter has benefitted from Peleg (1984), besides from the references in the text. Section 8.5 is based on Peleg and Procaccia (2007).

# Chapter 9
# Feasible elimination procedures

## 9.1 Motivation and summary

We have seen in Chapter 8 that a possible way to avoid the consequences of the Gibbard-Satterthwaite Theorem is to construct exactly and strongly consistent social choice functions. We recall that such functions ensure that the sincere outcome is an outcome of a strong Nash equilibrium for each profile of preferences of the voters. In this chapter and the next ones we investigate which effectivity functions (constitutions) admit exactly and strongly consistent social choice functions (voting procedures). This is a relevant question since voting is a basic characteristic of democratic societies. On a smaller scale, a society (cf. Definition 2.2.1) may be some committee and also then the existence of robust voting procedures is an important issue.

In this chapter we characterize all exactly and strongly consistent social choice functions which represent a fixed effectivity function from a family of anonymous effectivity functions (i.e., effectivity functions that do not distinguish between coalitions of the same size). From the results in the next chapter it will follow that every anonymous effectivity function without vetoers which has an exactly and strongly consistent representation, belongs to this family.

Our analysis of the class of exactly and strongly consistent social choice functions which represent this specific family of anonymous effectivity functions, is done in three stages. In Section 9.2 we show how to construct exactly and strongly consistent social choice functions by means of feasible elimination procedures relative to a system of blocking coefficients. In Section 9.3 we formulate the main problem of this chapter in terms of appropriate effectivity functions and in Section 9.4 we prove the characterization result.

B. Peleg and H. Peters, *Strategic Social Choice*, Studies in Choice and Welfare,
DOI 10.1007/978-3-642-13875-1_9, © Springer-Verlag Berlin Heidelberg 2010

## 9.2 Feasible elimination procedures

Let $A = \{a_1, \dots, a_m\}$ with $m \geq 2$ be a set of alternatives and let $N = \{1, \dots, n\}$ be a set of voters. We assume that $n + 1 \geq m$. Let the map $\beta : A \to N$ satisfy $\sum_{x \in A} \beta(x) = n+1$. Such a map $\beta$ may be used to define an anonymous effectivity function (see below) in which each alternative $x \in A$ can be blocked by each coalition of at least $\beta(x)$ voters – in other words, such a coalition is effective for $A \setminus \{x\}$. The assumption $\sum_{x \in A} \beta(x) = n + 1$ ensures that $N$ is effective for every $\{x\}$. For $B \in P_0(A)$ we denote $\beta(B) = \sum_{x \in B} \beta(x)$.

An effectivity function $E$ is *anonymous* if for all $B \subseteq A$ and $S, S' \subseteq N$, $B \in E(S)$ and $|S| = |S'|$ imply $B \in E(S')$.

With the help of the function $\beta$ we shall now construct a social choice correspondence (SCC) $M : L^N \to P_0(A)$, where $L$ denotes the set of *linear orderings* of $A$ (as in Section 8.2). Let $R^N \in L^N$. A *feasible elimination procedure* (f.e.p.) is a sequence $(x_1, C_1; \dots; x_{m-1}, C_{m-1}; x_m)$, where $C_s \in P_0(N)$ for $s = 1, \dots, m - 1$, such that:

$$C_t \cap C_s = \emptyset \text{ and } |C_s| = \beta(x_s) \text{ for all } s, t = 1, \dots, m-1, \ s \neq t. \tag{9.1}$$

$$A = \{x_1, \dots, x_m\}. \tag{9.2}$$

$$\{x_j, \dots, x_m\} R^i x_j \text{ for all } i \in C_j \text{ and } j = 1, \dots, m-1. \tag{9.3}$$

Observe that in a feasible elimination procedure the first element to be eliminated, $x_1$, is ranked last by (at least) the voters in the coalition $C_1$. This alternative, as well as the voters in $C_1$, are then eliminated from the profile. Next, $x_2$ is then ranked last by (at least) the voters in $C_2$, and $x_2$ as well as the voters in $C_2$ are eliminated from the profile. And so on and so forth. The last element, $x_m$, cannot be eliminated since the number of remaining voters is smaller than $\beta(x_m)$ by the condition $\sum_{x \in A} \beta(x) = n + 1$. Some further observations concerning f.e.p.'s are collected in the next remark.

*Remark 9.2.1.* For $R \in L$ and $j \in \{1, \dots, m\}$ let $t_j(R)$ denote the alternative ranked at position $j$, that is, $|\{x \in A \mid xRt_j(R)\}| = j$.

(a) Let $R^N \in L^N$. There must be an $x \in A$ and $S \subseteq N$ with $|S| = \beta(x)$ and $x = t_m(R^i)$ for all $i \in S$. For if not, then

$$n = \sum_{x \in A} \left|\{i \in N \mid x = t_m(R^i)\}\right| \leq \sum_{x \in A} (\beta(x) - 1) = n + 1 - m$$

which contradicts $m \geq 2$.

(b) Now suppose the alternatives $x_1, \dots, x_k$ with $1 \leq k \leq m-2$ have been eliminated. If in the remaining profile there would be no alternative $x$ and coalition $S$ with $|S| = \beta(x)$ and $x$ ranked last for all $i \in S$, then similarly as in (a) we would have

$$n - \sum_{j=1}^{k} \beta(x_j) \leq \sum_{x \in A \setminus \{x_1, \ldots, x_k\}} (\beta(x) - 1) = (n+1) - \sum_{j=1}^{k} \beta(x_j) - (m-k)$$

hence $m \leq k+1 \leq (m-2)+1 = m-1$, a contradiction. So by induction, it follows that for every $R^N \in L^N$ there exists a feasible elimination procedure.

(c) In fact, using the same argument as in (a) and (b) it follows that an f.e.p. may fail to exist if $\sum_{x \in A} \beta(x) > n+1$. Suppose $\sum_{x \in A} \beta(x) = n+\ell$ with $\ell \geq 1$. Then the inequality derived in (b) becomes $m \leq k+\ell \leq (m-2)+\ell$, which implies $\ell \geq 2$. So if we want an f.e.p. to exist for every profile of preferences we need $\ell = 1$.

(d) Let $R^N \in L^N$ and let $x \in A$. Suppose that there is an $S \subseteq N$ with $\beta(x) = |S|$ and $x = t_m(R^i)$ for all $i \in S$. Then $x$ is eliminated in each f.e.p. To see this, suppose there is an f.e.p. in which $x$ is not eliminated and let $y$ be the alternative eliminated last, say via coalition $T$. Then the finally left players form a coalition $S'$ containing $S$. We have $\beta(y) + \beta(x) = |T| + |S'| + 1$ by definition of $\beta$, but also $|T| + |S'| \geq \beta(y) + \beta(x)$, a contradiction.

To further illustrate the concept of an f.e.p. we consider an example.

*Example 9.2.2.* Let $A = \{a, b, c\}$ and $N = \{1, \ldots, 5\}$. Let $\beta(a) = \beta(b) = \beta(c) = 2$ and let $R^N$ be given by the following table.

| $R^1$ | $R^2$ | $R^3$ | $R^4$ | $R^5$ |
|---|---|---|---|---|
| $b$ | $c$ | $a$ | $c$ | $a$ |
| $c$ | $b$ | $b$ | $a$ | $c$ |
| $a$ | $a$ | $c$ | $b$ | $b$ |

Then there exist two f.e.p.'s: $(a, \{1,2\}; b, \{4,5\}; c)$ and $(b, \{4,5\}; a, \{1,2\}; c)$.

Let $R^N \in L^N$. The alternative $y \in A$ is *$R^N$-maximal* if there exists an f.e.p. $(x_1, C_1; \ldots; x_{m-1}, C_{m-1}; y)$ with respect to $R^N$. We denote

$$M(R^N) = \{x \in A \mid x \text{ is } R^N\text{-maximal}\}. \tag{9.4}$$

*Example 9.2.3.* Consider $A$, $N$, and $\beta$ as in Example 9.2.2. For the profile $R^N$ in this example we have $M(R^N) = \{c\}$. But $M$ may contain more than one alternative. Consider for instance the preference profile $Q^N \in L^N$ given by the following table.

| $Q^1$ | $Q^2$ | $Q^3$ | $Q^4$ | $Q^5$ |
|---|---|---|---|---|
| $b$ | $c$ | $b$ | $c$ | $b$ |
| $c$ | $b$ | $c$ | $a$ | $c$ |
| $a$ | $a$ | $a$ | $b$ | $a$ |

Then $M(Q^N) = \{b, c\}$.

*Remark 9.2.4.* The SCC $M : L^N \to P_0(A)$ is obviously *Paretian*, i.e., $M(R^N)$ contains only Pareto optimal alternatives for every $R^N \in L^N$ (cf. Definition 3.3.7). It is also anonymous: an SCC $H : L^N \to P_0(A)$ is *anonymous* if for all permutations $\pi : N \to N$ and all $R^N \in L^N$ we have $H(R^1, \ldots, R^n) = H(R^{\pi(1)}, \ldots, R^{\pi(n)})$.

Recall (see Section 8.2) that a social choice function (SCF) is a map $F: L^N \to A$. An SCF $F$ may be considered as a special case of an SCC by considering the SCC $H_F(R^N) = \{F(R^N)\}$. Then, $F$ is *Paretian* (*anonymous*) if $H_F$ is Paretian (anonymous). Our main concern in this chapter are SCF's that are selections from $M$. We first observe that if an SCF $F$ is a selection from $M$, that is, $F(R^N) \in M(R^N)$ for all $R^N \in L^N$, then $F$ is Paretian. Second, there exist anonymous selections from $M$ – for instance, select the maximal element of $M(R^N)$ according to a fixed linear ordering in $L$.

We have argued in Chapter 8 that one possibility of avoiding the Gibbard-Satterthwaite Theorem is to construct exactly and strongly consistent (ESC) social choice functions. We shall now prove that every selection from $M$ is ESC. For convenience of the reader we replicate the precise definition. Note that an SCF $F$ can be seen as a game form in which each player has strategy set $L$ and $F$ is the outcome function. Then $(F, R^N)$ is a game with player preferences given by $R^N$. In a strong (Nash) equilibrium, no coalition can gain by deviating, cf. Definition 5.2.1.

**Definition 9.2.5.** A social choice function $F: L^N \to A$ is *exactly and strongly consistent* if for every $R^N \in L^N$ the game $(F, R^N)$ has a strong equilibrium $Q^N \in L^N$ such that $F(Q^N) = F(R^N)$.

The main result of this section is as follows.

**Theorem 9.2.6.** *Let $A = \{a_1, \ldots, a_m\}$, $m \geq 2$, $N = \{1, \ldots, n\}$, $n + 1 \geq m$, and $\beta: A \to N$ such that $\sum_{x \in A} \beta(x) = n + 1$. Then every selection from $M$ is exactly and strongly consistent.*

*Proof.* Let an SCF $F$ be a selection from $M$ and let $R^N \in L^N$. Denote $y = F(R^N)$. Since $y \in M(R^N)$ there exists an f.e.p. $(x_1, C_1; \ldots; x_{m-1}, C_{m-1}; y)$ with respect to $R^N$. Take $Q^N \in L^N$ with $t_m(Q^\ell) = x_j$ for all $j = 1, \ldots, m-1$ and $\ell \in C_j$. Then $M(Q^N) = \{y\}$ by Remark 9.2.1, parts (b) and (d), so $F(Q^N) = y = F(R^N)$. Also, $Q^N$ is a strong equilibrium of $(F, R^N)$. Indeed, assume on the contrary that there exist $S \subseteq N$ and $P^S \in L^S$ such that $F(P^S, Q^{N \setminus S}) = z \neq y$ and $zR^iy$ for all $i \in S$. Then $z = x_j$ for some $1 \leq j \leq m-1$. By (9.3), $yR^iz$ for all $i \in C_j$, hence $S \cap C_j = \emptyset$. Since $|C_j| = \beta(z)$ and $z = t_m(Q^\ell)$ for all $\ell \in C_j$, Remark 9.2.1 part (d) implies $z \notin M(P^S, Q^{N \setminus S})$. So $F(P^S, Q^{N \setminus S}) \neq z$, which is the desired contradiction. □

## 9.3 Maximal alternatives and effectivity functions

In this section we again assume $n+1 \geq m$, and consider a function $\beta: A \to N$ with $\sum_{x \in A} \beta(x) = n + 1$ and the associated social choice correspondence $M$ assigning maximal alternatives, i.e., alternatives resulting from feasible elimination procedures. We first characterize $M$ in terms of $\beta$ through the use of effectivity functions. To this end we define an effectivity function $E_\beta : P(N) \to P(P_0(A))$ by $E(\emptyset) = \emptyset$ and

$$B \in E_\beta(S) \Leftrightarrow |S| \geq \sum_{x \notin B} \beta(x) = \beta(A \setminus B), \text{ for all } S \in P_0(N) \text{ and } B \in P_0(A).$$

It is easy to check that $E_\beta$ is indeed an effectivity function (Definition 2.3.1).

We claim that $E_\beta$ is also superadditive and maximal. To prove superadditivity, let $B_i \in E(S_i)$, $i = 1, 2$, and $S_1 \cap S_2 = \emptyset$. Then

$$\begin{aligned} |S_1 \cup S_2| &= |S_1| + |S_2| \\ &\geq \beta(A \setminus B_1) + \beta(A \setminus B_2) \\ &\geq \beta\left((A \setminus B_1) \cup (A \setminus B_2)\right) \\ &= \beta\left(A \setminus (B_1 \cap B_2)\right), \end{aligned}$$

so that $B_1 \cap B_2 \in E(S_1 \cup S_2)$.

To prove maximality, suppose $S \neq \emptyset, N$ and $\emptyset \neq B \notin E_\beta(S)$. Then $|S| < \beta(A \setminus B) = n + 1 - \beta(B)$, so that $n - |S| > \beta(B) - 1$. Hence, $n - |S| \geq \beta(B)$ so that $A \setminus B \in E_\beta(N \setminus S)$.

Let $E^M$ be the effectivity function induced by the social choice correspondence $M$ (see Section 8.5 for the exact definition). Recall that $E^M$ is superadditive. It turns out that $E^M$ and $E_\beta$ are identical. We first formulate a general observation about effectivity functions.

**Lemma 9.3.1.** *Let $E, E' : P(N) \to P(P_0(A))$ be effectivity functions with $E(S) \subseteq E'(S)$ for all $S \in P_0(N)$, let $E$ be maximal and let $E'$ be superadditive. Then $E = E'$.*

*Proof.* Let $S \in P_0(N)$ and $B \in E'(S)$, and suppose for contradiction that $B \notin E(S)$. Then, by maximality of $E$, $A \setminus B \in E(N \setminus S)$. Hence $A \setminus B \in E'(N \setminus S)$, but this contradicts superadditivity of $E'$, since also $B \in E'(S)$. Hence we must have $B \in E(S)$. □

**Lemma 9.3.2.** $E^M = E_\beta$.

*Proof.* We first prove $E_\beta \subseteq E^M$. Let $S \in P_0(N)$ and $B \in E_\beta(S)$. Then $|S| \geq \beta(A \setminus B)$. Let $A \setminus B = \{x_1, \ldots, x_k\}$ and let $S_1, \ldots, S_k$ be a partition of $S$ such that $|S_j| \geq \beta(x_j)$ for all $j = 1, \ldots, k$. Choose $R^S \in L^S$ such that $t_m(R^\ell) = x_j$ for all $\ell \in S_j$ and $j = 1, \ldots, k$. Then, by Remark 9.2.1 part (d), $M(R^S, Q^{N \setminus S}) \subseteq B$ for all $Q^{N \setminus S} \in L^{N \setminus S}$. So $B \in E^M(S)$.

The lemma now follows from the first part of the proof and Lemma 9.3.1, since $E_\beta$ is maximal and $E^M$ is superadditive. □

Since a social choice function $F$ is a special case of a social choice correspondence the definition in Section 8.5 also defines a (superadditive) effectivity function $E^F$. Theorem 9.2.6 and Lemma 9.3.2 now yield the following interesting corollary. First we make a simple but useful observation.

*Remark 9.3.3.* Let $F, F' : L^N \to P_0(A)$ be two social choice correspondences with $F(R^N) \subseteq F'(R^N)$ for all $R^N \in L^N$. Then $E^F(S) \supseteq E^{F'}(S)$ for all $S \in P_0(N)$. To see this, let $S \in P_0(N)$ and $B \in E^{F'}(S)$. Then there is $Q^S \in L^S$ such that $F'(Q^S, Q^{N\setminus S}) \subseteq B$ for all $Q^{N\setminus S} \in L^{N\setminus S}$. Hence $F(Q^S, Q^{N\setminus S}) \subseteq B$ for all $Q^{N\setminus S} \in L^{N\setminus S}$, so that $B \in E^F(S)$.

**Corollary 9.3.4.** *If a social choice function $F$ is a selection from $M$, then (i) $F$ is exactly and strongly consistent and (ii) $E^F = E_\beta$.*

*Proof.* Clearly, we only have to prove (ii). Since $E^M = E_\beta$ by Lemma 9.3.2 we have in particular that $E^M$ is maximal. Also, $E^F$ is superadditive. Hence, to prove that $E^F = E^M$, it is by Lemma 9.3.1 sufficient to prove that $E^M(S) \subseteq E^F(S)$ for all $S \in P_0(N)$, but this follows from Remark 9.3.3. □

It turns out that the converse of Corollary 9.3.4 is true as well.

**Theorem 9.3.5.** *If a social choice function $F$ is exactly and strongly consistent and $E^F = E_\beta$, then $F$ is a selection from $M$.*

Theorem 9.3.5 is a direct corollary of the following characterization of $M$, which states that $M$ coincides with the core of $E_\beta$.

**Theorem 9.3.6.** $M(R^N) = C(E_\beta, R^N)$ *for all* $R^N \in L^N$.

We will provide a proof of Theorem 9.3.6 in the next section. To see why Theorem 9.3.5 follows, note that an SCF $F$ satisfying the conditions in this theorem must be a selection from the core of its associated effectivity function $E^F$. This follows from Proposition 5.2.6. Thus, $F(R^N) \in C(E^F, R^N) = C(E_\beta, R^N) = M(R^N)$ for all $R^N \in L^N$, and Theorem 9.3.5 follows.

*Remark 9.3.7.* Since $M(R^N) \neq \emptyset$ for all $R^N \in L^N$ by Remark 9.2.1 part (b), Theorem 9.3.6 implies in particular that $E_\beta = E^M$ is stable. Also, since $C(E_\beta, \cdot)$ is a Maskin monotonic SCC (see Remark 3.7.3), $M$ is Maskin monotonic.

*Remark 9.3.8.* Maskin monotonicity can be weakened as follows. Let $R^N \in L^N$ and $x \in A$. Call $R_1^N$ an *$x$-improvement* of $R^N$ if (i) for all $a \in A \setminus \{x\}$ and all $i \in N$, $xR^ia \Rightarrow xR_1^ia$ and (ii) for all $a, b \in A \setminus \{x\}$ and all $i \in N$, $aR^ib \Leftrightarrow aR_1^ib$. Then a social choice correspondence $H : L^N \to P_0(A)$ is *monotonic* if for all $R^N \in L^N$, all $x \in H(R^N)$ and all $x$-improvements $R_1^N \in L^N$ we have (i) $x \in H(R_1^N)$ and (ii) $H(R_1^N) \subseteq H(R^N)$.

Maskin monotonicity of $H$ implies monotonicity. This was proved in Peleg, 1984, Section 2.3, but for the sake of completeness we provide a proof here. Note that it is sufficient to consider the case where $R_1^N$ arises from $R^N$ by switching $x$ with the alternative ranked right above $x$, say $y$, in the preference of just one player. By Maskin monotonicity, $x \in H(R_1^N)$. Suppose there is an alternative $z \in A \setminus \{x\}$ with $z \in H(R_1^N)$. By Maskin monotonicity again, now applied to $z$, we obtain $z \in H(R^N)$. Thus, $H(R_1^N) \subseteq H(R^N)$.

Clearly, the core of a stable effectivity function is a monotonic SCC. A social choice function $F : L^N \to A$ is *monotonic* if the associated SCC $H_F(\cdot) = \{F(\cdot)\}$ is monotonic. It is not hard to see that if $H$ is a monotonic SCC, then there exists a monotonic SCF $F$ which is a selection from $H$: simply fix a linear order in $L$ and let $F(R^N)$ be the maximal element of $H(R^N)$ according to this linear order. In particular, in this way we can construct an anonymous, Paretian and monotonic selection from $M$.

## 9.4 A proof of Theorem 9.3.6

This section is completely devoted to proving Theorem 9.3.6.

First, let $R^N \in L^N$ and $x \in M(R^N)$. Choose a selection $F$ from $M$ such that $F(R^N) = x$. Then, by Corollary 9.3.4, $E^F = E_\beta$ and by the same corollary and Proposition 5.2.6, $F(R^N) \in C(E^F, R^N)$. Thus, $M(R^N) \subseteq C(E_\beta, R^N)$.

The converse inclusion $M(R^N) \supseteq C(E_\beta, R^N)$ for all $R^N \in L^N$ is proved by induction on $m$. For $m = 2$ and $R^N \in L^N$, obviously, both $M(R^N)$ and $C(E_\beta, R^N)$ coincide with the same singleton. Henceforth, $m \geq 3$. Let $R^N \in L^N$ and let $x \in C(E_\beta, R^N)$. Denote $A \setminus \{x\} = \{y_1, \ldots, y_{m-1}\}$ and let $S(y) = \{i \in N \mid xR^iy\}$ for all $y \in \{y_1, \ldots, y_{m-1}\}$. Since $x \in C(E_\beta, R^N)$, we have for every $B \subseteq A \setminus \{x\}$:

$$|\cap_{y \in B} N \setminus S(y)| = \big|\{i \in N \mid yR^ix \text{ for all } y \in B\big| < \beta(A \setminus B) = n + 1 - \beta(B).$$

Hence $n - |\cup_{y \in B} S(y)| < n + 1 - \beta(B)$, or $|\cup_{y \in B} S(y)| \geq \beta(B)$. It is easy to see that this argument can be reversed, so that:

$$x \in C(E_\beta, R^N) \Leftrightarrow |\cup_{y \in B} S(y)| \geq \beta(B) \text{ for all } B \subseteq A \setminus \{x\}. \tag{9.5}$$

We first consider the following case:

$$\text{There exists } B_0 \subseteq A \setminus \{x\} \text{ such that } 1 \leq |B_0| \leq m - 2 \text{ and } |\cup_{y \in B_0} S(y)| = \beta(B_0). \tag{9.6}$$

We now decompose the problem into two (non-disjoint) subproblems:[1]

(1) $N_1 = \cup_{y \in B_0} S(y)$, $A_1 = B_0 \cup \{x\}$, $\beta_1(y) = \beta(y)$ for all $y \in B_0$, $\beta_1(x) = 1$, and $R_1^{N_1} = R^{N_1}{}_{|A_1}$.

(2) $N_2 = N \setminus N_1$, $A_2 = A \setminus B_0$, $\beta_2(y) = \beta(y)$ for all $y \in A_2$, and $R_2^{N_2} = R^{N_2}{}_{|A_2}$.

By (9.5), $x \in C(E_{\beta_1}, R_1^{N_1}) \cap C(E_{\beta_2}, R_2^{N_2})$. Hence by the induction hypothesis there exists an f.e.p. $(z_1, C_1; \ldots; z_k, C_k; x)$ for subproblem (1), where $k =$

[1] By $R^S{}_{|B}$ we denote the restriction of the profile $R^S$ for a coalition $S$ to a set of alternatives $B$.

$|B_0|$, and an f.e.p. $(u_1, D_1; \ldots; u_{m-1-k}, D_{m-1-k}; x)$ for subproblem (2). Since $B_0 R^i x$ for all $i \in N_2$ by definition of $N_1$ and $N_2$, it follows that

$$(u_1, D_1; \ldots; u_{m-1-k}, D_{m-1-k}; z_1, C_1; \ldots; z_k, C_k; x)$$

is an f.e.p. for the original problem. Thus, in case (9.6), $x \in M(R^N)$.

We now consider the complementary case, i.e.,

$$|\cup_{y \in B} S(y)| > \beta(B) \text{ for all } B \subseteq A \setminus \{x\} \text{ with } 1 \leq B \leq m-2. \tag{9.7}$$

Suppose there is an $\ell \in N$ with $x = t_j(R^\ell)$ for some $j \leq m-2$, and denote $\widehat{y} = t_{j+1}(R^\ell)$. We switch $x$ and $\widehat{y}$ in player $\ell$'s preference to obtain a new preference $\widehat{R}^\ell$ and a new preference profile $\widehat{R}^N = (R^1, \ldots, R^{\ell-1}, \widehat{R}^\ell, R^{\ell+1}, \ldots, R^N)$ that still satisfies the right hand side of (9.5), hence, $x \in C(E_\beta, \widehat{R}^N)$ – note, in particular, that the right hand side of (9.5) still holds for $B = A \setminus \{x\}$ since $x \neq t_m(\widehat{R}^\ell)$ and, thus, $\ell \in S(t_m(\widehat{R}^\ell))$. If (9.6) holds for $\widehat{R}^N$, then by the first case, $x \in M(\widehat{R}^N)$. Thus, by monotonicity (cf. Remark 9.3.8), $x \in M(R^N)$. If (9.6) does not hold for $\widehat{R}^N$, then we repeat this step for some player $\ell' \in N$ with $x = t_j(\widehat{R}^{\ell'})$ for some $j \leq m-2$, and so on, until either (9.6) is satisfied or there is no player left with $x$ not ranked at the last or before last position.

In the latter case, we have a profile, say $\widetilde{R}^N$, with still $x \in C(E_\beta, \widetilde{R}^N)$, and $x = t_{m-1}(\widetilde{R}^i)$ or $x = t_m(\widetilde{R}^i)$ for all $i \in N$. Since $|S(y_j)| \geq \beta(y_j)$ for all $j = 1, \ldots, m-1$ by (9.5) (for $\widetilde{R}^N$), we can take sets $S_j \subseteq S(y_j)$ for all $j = 1, \ldots, m-1$ such that $(y_1, S_1; \ldots; y_{m-1}, S_{m-1}; x)$ is an f.e.p. for $\widetilde{R}^N$. Thus, $x \in M(\widetilde{R}^N)$ and by monotonicity again, $x \in M(R^N)$. □

## 9.5 Notes and comments

Section 9.2 is based on Peleg (1978a) and on Oren (1981). In particular, the definition of a feasible elimination procedure and the proof of Theorem 9.2.6 appeared for the first time in Peleg (1978a).

Theorem 9.3.6 is due to Polishchuk (1978). It appeared in print first in Peleg (1984). The proof presented above is new and much more transparent than the proof of Theorem 5.4.2 in Peleg (1984). This new proof has benefitted from the analysis in Peleg and Peters (2006), see also Chapter 11.

# Chapter 10
# Exactly and strongly consistent representations of effectivity functions

## 10.1 Motivation and summary

As argued in the introductory section of the previous chapter it is important to find robust voting procedures for constitutions. In our approach, 'robust' means 'exactly and strongly consistent' (ESC): the game induced by the social choice function should have a strong equilibrium for each profile of preferences which, moreover, results in the same outcome as truthful voting.

In this chapter we extend the study of ESC representations to general effectivity functions. We start, in Section 10.2, by generalizing the definition of a feasible elimination procedure to arbitrary effectivity functions: the generalization uses blocking coalitions instead of blocking coefficients. An effectivity function will be called 'elimination stable' if it admits at least one feasible elimination procedure (f.e.p.) for each profile of linear orderings (strict preferences). We show that if the effectivity function is maximal, stable, and elimination stable, then it has an ESC representation. In fact, each selection from the maximal alternatives – alternatives that result from an f.e.p. – is an ESC representation.

Thus, we conclude from Section 10.2 that the key concept associated with the existence of ESC representations is elimination stability. Following Holzman (1986b) we determine sufficient conditions, denoted as $D(0), \ldots, D(m-2)$, for the existence of f.e.p.'s for every profile of preferences. These conditions are somewhat technical but arise naturally if one seeks to construct f.e.p.'s. The main result of Section 10.3 is Theorem 10.3.2.

A natural follow-up question then is whether these conditions are also necessary for the existence of an ESC representation. Surprisingly, there is a partial converse. Suppose that $E$ is an effectivity function without individual rights, that is, singleton coalitions are only effective for the whole set $A$. Then $E$ has an ESC representation only if it satisfies $D(0), \ldots, D(m-2)$. See Theorems 10.4.5 and 10.5.2. Thus, an effectivity function without individual rights has an ESC representation if and only if it has such a representation

B. Peleg and H. Peters, *Strategic Social Choice*, Studies in Choice and Welfare,
DOI 10.1007/978-3-642-13875-1_10, © Springer-Verlag Berlin Heidelberg 2010

by feasible elimination procedures, that is, by a selection from its maximal alternatives.

## 10.2 Feasible elimination procedures revisited

We start by extending the definition of a feasible elimination procedure, introduced in Chapter 9, to general, and not necessarily anonymous, effectivity functions. Let $E : P(N) \to P(P_0(N))$ be an effectivity function and let $R^N \in L^N$. A *feasible elimination procedure* (f.e.p.) with respect to $E$ and $R^N$ is a sequence $(x_1, S_1; \ldots; x_{m-1}, S_{m-1}; x_m)$ such that:

$$A = \{x_1, \ldots, x_m\},\ S_1, \ldots, S_{m-1} \in P_0(N). \tag{10.1}$$

$$S_i \cap S_j = \emptyset \text{ for all } i, j = 1, \ldots, m-1,\ i \neq j. \tag{10.2}$$

$$A \setminus \{x_j\} \in E(S_j) \text{ for all } j = 1, \ldots, m-1. \tag{10.3}$$

$$\{x_j, \ldots, x_m\} R^{S_j} x_j \text{ for all } j = 1, \ldots, m-1. \tag{10.4}$$

The interpretation of such an f.e.p. is similar to the interpretation in Chapter 9. Coalition $S_1$ is effective for $A \setminus \{x_1\}$ and, moreover, prefers $A \setminus \{x_1\}$ over $x_1$. Then the alternative $x_1$ and the players in $S_1$ are removed from the profile, and the procedure is repeated for $S_2$ and $x_2$; and so on and so forth. In the end only $x_m$ is left.

An alternative $y \in A$ is $R^N$*-maximal* if there exists an f.e.p. $(x_1, S_1; \ldots; x_{m-1}, S_{m-1}; x_m)$ with respect to $E$ and $R^N$ such that $x_m = y$. We denote $M(E, R^N) = \{y \in A \mid y \text{ is } R^N\text{-maximal}\}$. The effectivity function $E$ is *elimination stable* if $M(E, R^N) \neq \emptyset$ for all $R^N \in L^N$.

Recall that an effectivity function $E$ is stable if its core $C(E, R^N)$ is nonempty for all $R^N \in L^N$. We say that $E$ has an *exactly and strongly consistent* (ESC) representation if there is an ESC social choice function $F$ (cf. Definition 9.2.5) with $E = E^F$. Thus, for every profile $R^N$ of preferences the game $(F, R^N)$ has a strong Nash equilibrium $Q^N$ with $F(Q^N) = F(R^N)$.

If $E$ has an ESC representation $F$ then of course $F$ is a strongly consistent representation and thus, in particular, $E$ is maximal and stable – this follows from Propositions 5.2.4 and 5.2.6. Since we aim at finding ESC representations of $E$ we shall often assume stability and maximality of $E$ in the remainder of this chapter.

The first result shows that maximal alternatives of stable effectivity functions are in the core.

**Theorem 10.2.1.** *Let $E$ be stable. Then $M(E, R^N) \subseteq C(E, R^N)$ for all $R^N \in L^N$.*

*Proof.* Suppose $x \in M(E, R^N)$ and let $(x_1, S_1; \ldots; x_{m-1}, S_{m-1}; x)$ be an f.e.p. Define $Q^N \in L^N$ by $t_m(Q^i) = x_j$ for all $i \in S_j$ and $j = 1, \ldots, m-1$,

${Q^i}_{|A\setminus\{x_j\}} = {R^i}_{|A\setminus\{x_j\}}$ for all $i \in S_j$ and $j = 1, \ldots, m-1$, and $Q^i = R^i$ for all $i \in N \setminus \cup_{j=1}^{m-1} S_j$. Then $x_j \notin C(E, Q^N)$ for all $j = 1, \ldots, m-1$. Hence, by stability of $E$, $\{x\} = C(E, Q^N)$. Since $xR^\ell x_j$ for all $j = 1, \ldots, m-2$ and $\ell \in S_j$, Maskin monotonicity of $C(E, \cdot)$ implies $x \in C(E, R^N)$. □

The next result generalizes Theorem 9.2.6 to elimination stable effectivity functions.

**Theorem 10.2.2.** *Let $E$ be stable. If $E$ is elimination stable and the social choice function $F : L^N \to A$ is a selection from $M(E, \cdot)$, then $F$ is exactly and strongly consistent.*

*Proof.* Let $E$ and $F$ be as in the statement of the theorem, $R^N \in L^N$, and denote $x = F(R^N) \in M(E, R^N)$. Let $(x_1, S_1; \ldots; x_{m-1}, S_{m-1}; x)$ be an f.e.p., and define $Q^N \in L^N$ as in the proof of Theorem 10.2.1. Just like there, $\{x\} = C(E, Q^N)$, so by Theorem 10.2.1, $\{x\} = M(E, Q^N)$. Thus, $F(Q^N) = F(R^N) = x$. We claim that $Q^N$ is a strong equilibrium of $(F, R^N)$. Indeed, assume on the contrary that there exists $S \in P_0(N)$, $P^S \in L^S$, and $y \in A$, such that $F(P^S, Q^{N \setminus S}) = y \neq x$, and $yR^i x$ for all $i \in S$. Let $y = x_j$. Then, by (10.4), $S \cap S_j = \emptyset$. Hence $y \notin C(E, (P^S, Q^{N\setminus S}))$. So by Theorem 10.2.1, $y \notin M(E, (P^S, Q^{N\setminus S}))$ and, thus, $y \neq F(P^S, Q^{N\setminus S})$, a contradiction. □

We proceed with the following observation.

**Lemma 10.2.3.** *Let $E$ be maximal, stable, and elimination stable, and denote $M(\cdot) = M(E, \cdot)$. Then $E^M = E$.*

*Proof.* Since $E$ is maximal and $E^M$ is superadditive, it is by Lemma 9.3.1 sufficient to prove that $E(S) \subseteq E^M(S)$ for every $S \in P_0(N)$. Denote $C(\cdot) = C(E, \cdot)$. By Theorem 10.2.1, $M(R^N) \subseteq C(R^N)$ for all $R^N \in L^N$, so by Remark 9.3.3, $E^C(S) \subseteq E^M(S)$ for every $S$. Hence, it is sufficient to prove that $E(S) \subseteq E^C(S)$ for every $S \in P_0(N)$. Let $S \in P_0(N)$ and $B \in E(S)$. Consider $Q^N \in L^N$ such that $B\, Q^i\, (A \setminus B)$ for all $i \in S$. Then, clearly, $C(Q^N) \subseteq B$. Hence, $B \in E^C(S)$. □

This leads to the main result of this section.

**Corollary 10.2.4.** *Let $E$ be maximal, stable, and elimination stable. Let the social choice function $F$ be a selection from $M(\cdot) = M(E, \cdot)$. Then $F$ is an exactly and strongly consistent representation of $E$.*

*Proof.* By Theorem 10.2.2 it is sufficient to prove that $E = E^F$. Since $E$ is maximal and $E^F$ is superadditive, it is by Lemma 9.3.1 sufficient to prove that $E(S) \subseteq E^F(S)$ for all $S$. By Remark 9.3.3 we have $E^M(S) \subseteq E^F(S)$ for all $S$. Hence, by Lemma 10.2.3, $E(S) = E^M(S) \subseteq E^F(S)$ for all $S$. □

## 10.3 Sufficient conditions for elimination stability

We have seen in Section 10.2 that elimination stability of an effectivity function, combined with the necessary conditions of stability and maximality, is a sufficient condition for the existence of an exactly and strongly consistent representation (Corollary 10.2.4). We shall see that, under an additional condition on the effectivity function, elimination stability is also necessary for this (Theorem 10.5.2, Corollary 10.5.3). In this section we establish sufficient conditions for elimination stability.

Let $E : P(N) \to P(P_0(A))$ be an effectivity function. We shall define a sequence of conditions $D(k)$, $k = 0, \dots, m-2$, on $E$ that guarantee the nonemptiness of $M(E, R^N)$ for every $R^N \in L^N$ – that is, elimination stability of $E$.

In the sequel we are going to use the concept of a 'generalized partition': a *generalized partition* (g-partition) of a set is a partition where some of the elements may be empty.

A coalition $S \subseteq N$ is *blocking* for $x \in A$ if $A \setminus \{x\} \in E(S)$. Denote $\mathcal{B}(x) = \{S \subseteq N \mid S \text{ is blocking for } x\}$. If $S \in \mathcal{B}(x)$ but no proper subset of $S$ is in $\mathcal{B}(x)$, then $S$ is *minimal blocking*. The set of minimal blocking coalitions for $x$ is denoted by $\mathcal{B}_m(x)$.

**Definition 10.3.1.** Let $E$ be an effectivity function and let $k$ be an integer, $0 \le k \le m-2$. We say that $E$ *satisfies condition* $D(k)$ if there exist *no* enumeration $x_1, \dots, x_m$ of $A$ and g-partition $S_1, \dots, S_m$ of $N$ such that $S_j \in \mathcal{B}_m(x_j)$ for $j = 1, \dots, k$ and $S_j \notin \mathcal{B}(x_j)$ for $j = k+1, \dots, m$.

These conditions $D(k)$ for $k = 0, \dots, m-2$ will enable us to construct, inductively, a feasible elimination procedure for every $R^N$. This is made precise in the proof of the following theorem.

**Theorem 10.3.2.** *Let $A = \{a_1, \dots, a_m\}$ and let $E$ be an effectivity function satisfying $D(k)$ for $k = 0, \dots, m-2$. Then $E$ is elimination stable.*

*Proof.* Let $R^N \in L^N$. We construct an f.e.p. inductively. In the first step let $\widehat{S}_j = \{i \in N \mid t_m(R^i) = a_j\}$ for $j = 1, \dots, m$. Then $\widehat{S}_1, \dots, \widehat{S}_m$ is a g-partition. By $D(0)$ there exists $j_0$ such that $\widehat{S}_{j_0} \in \mathcal{B}(a_{j_0})$. Denote $x_1 = a_{j_0}$ and choose $S_1 \subseteq \widehat{S}_{j_0}$, $S_1 \in \mathcal{B}_m(x_1)$. Then we let $(x_1, S_1)$ be the first component of the f.e.p. to be constructed.

Assume now that $(x_1, S_1; \dots; x_k, S_k)$, $k \ge 1$, is an initial segment of our f.e.p. That is, $S_j \in \mathcal{B}_m(x_j)$ for $j = 1, \dots, k$; $S_i \cap S_j = \emptyset$ for $i \ne j$; and $A \setminus \{x_1, \dots, x_j\} R^{S_j} x_j$ for $j = 1, \dots, k$. Then, for each $x \in A \setminus \{x_1, \dots, x_k\}$ let $S_x = \{s \in N \setminus \cup_{i=1}^k S_i \mid \; yR^s x \text{ for all } y \in A \setminus \{x_1, \dots, x_k\}\}$. By condition $D(k)$ there is an $x \in A \setminus \{x_1, \dots, x_k\}$ such that $S_x \in \mathcal{B}(x)$. Set $x = x_{k+1}$ and choose $S_{k+1} \subseteq S_x$ such that $S_{k+1} \in \mathcal{B}_m(x_{k+1})$. Then the proof is complete after $m-1$ steps. □

## 10.4 Necessary conditions for the existence of ESC representations

In this section we formulate a sequence of necessary conditions $D^*(k)$ for the existence of an exactly and strongly consistent representation of an effectivity function $E$, under an additional assumption on $E$. In Section 10.5 we show that these conditions imply the sufficient conditions $D(k)$ of Section 10.3.

The additional assumption on $E$ is the following.

$$E(\{i\}) = \{A\} \text{ for all } i \in N. \tag{10.5}$$

This assumption can be expressed as saying that $E$ has no *vetoers.*

The announced conditions $D^*(k)$ are obtained by strengthening the conditions $D(k)$ in order to deal with *sets* of alternatives. We say that a coalition $S \subseteq N$ *blocks* $B \subseteq A$ if $A \setminus B \in E(S)$. We denote by $\mathcal{B}(B)$ the set of coalitions that block $B$. We start by formulating $D^*(0)$.

**Definition 10.4.1.** Let $E$ be an effectivity function. We say that $E$ *satisfies condition* $D^*(0)$ if there exists *no* partition $C_1, \ldots, C_p$ of $A$ and no g-partition $S_1, \ldots, S_p$ of $N$, $p \geq 2$, such that $S_i \notin \mathcal{B}(C_i)$ for all $i = 1, \ldots, p$.

The following result implies that $D^*(0)$ is a necessary condition for the existence of an ESC representation of an effectivity function.

**Lemma 10.4.2.** *Let $E$ be a stable and maximal effectivity function. Then $E$ satisfies $D^*(0)$.*

*Proof.* Suppose, on the contrary, that there exist a partition $C_1, \ldots, C_p$ of $A$ and a g-partition $S_1, \ldots, S_p$ of $N$, $p \geq 2$, that violate $D^*(0)$, that is, $S_i \notin \mathcal{B}(C_i)$ for all $i = 1, \ldots, p$. Consider a profile $R^N$ as in the following table:

| $S_1$ | $S_2$ | $\cdots$ | $S_p$ |
|---|---|---|---|
| $C_2$ | $C_3$ | $\cdots$ | $C_1$ |
| $C_3$ | $C_4$ | $\cdots$ | $C_2$ |
| $\vdots$ | $\vdots$ | $\vdots$ | $\vdots$ |
| $C_p$ | $C_1$ | $\cdots$ | $C_{p-1}$ |
| $C_1$ | $C_2$ | $\cdots$ | $C_p$ |

where the ordering of the alternatives inside the sets $C_i$ does not matter. Now $S_1 \notin \mathcal{B}(C_1)$ means $A \setminus C_1 \notin E(S_1)$. By maximality, $C_1 \in E(N \setminus S_1)$. Since $C_1 \, R^{N \setminus S_1} \, C_2$, we have $C_2 \cap C(E, R^N) = \emptyset$. Similarly one shows $C_i \cap C(E, R^N) = \emptyset$ for all $i = 1, \ldots, p$. Hence, $C(E, R^N) = \emptyset$, which contradicts stability of $E$. □

We next formulate the more involved conditions $D^*(1), \ldots, D^*(m-2)$.

**Definition 10.4.3.** Let $E$ be an effectivity function and let $1 \leq k \leq m-2$. We say that $E$ *satisfies condition* $D^*(k)$ if there exist *no* $x_1, \ldots, x_k \in A$, $C_1, C_2 \in P(A)$, and $S_1, \ldots, S_k, T_1, T_2 \in P(N)$ such that

$$\{x_1\}, \ldots, \{x_k\}, C_1, C_2 \text{ is a partition of } A; \tag{10.6}$$

$$S_1, \ldots, S_k, T_1, T_2 \text{ is a g-partition of } N; \tag{10.7}$$

$$S_i \in \mathcal{B}_m(x_i) \text{ for } i = 1, \ldots, k; \text{ and} \tag{10.8}$$

$$T_i \notin \mathcal{B}(C_i) \text{ for } i = 1, 2. \tag{10.9}$$

**Lemma 10.4.4.** *Let the effectivity function $E$ satisfy* (10.5) *and let $E$ have an ESC representation. Then $E$ satisfies $D^*(1)$.*

*Proof.* Suppose, on the contrary, that $D^*(1)$ is violated and let $\{x\}, C_1, C_2$ be a partition of $A$ and $S, T_1, T_2$ a g-partition of $N$, such that $S \in \mathcal{B}_m(x)$ and $T_i \notin \mathcal{B}(C_i)$ for $i = 1, 2$.

By (10.5), $|S| > 1$. Suppose $S$ were equal to $N$. Let $i \in N$. Then $N \setminus \{i\} \notin \mathcal{B}(x)$ since $N \in \mathcal{B}_m(x)$. Hence $A \setminus \{x\} \notin E(N \setminus \{i\})$. By maximality of $E$ (which follows since $E$ has an ESC representation by assumption) this implies $\{x\} \in E(\{i\})$, which violates (10.5). Hence, $1 < |S| < n$.

Let $S^{(1)}, S^{(2)}$ be a partition of $S$ into two non-empty coalitions and consider the following profile $R^N$:

| $S^{(1)}$ | $S^{(2)}$ | $T_1$ | $T_2$ |
|---|---|---|---|
| $C_2$ | $C_1$ | $x$ | $x$ |
| $C_1$ | $C_2$ | $C_2$ | $C_1$ |
| $x$ | $x$ | $C_1$ | $C_2$ |

Let $F$ be an ESC representation of $E$. Since $E^F = E$, $F(R^N) \in C(E^F, R^N)$ (by Proposition 5.2.6), and $x \notin C(E, R^N)$ (as $S \in \mathcal{B}_m(x)$), we have $F(R^N) \neq x$. Without loss of generality, $F(R^N) \in C_1$. Let $Q^N \in L^N$ be a strong equilibrium of $(F, R^N)$ with $F(Q^N) = F(R^N)$. We distinguish two cases.

*Case 1.* There exist $i \in S$ and $y \in A \setminus \{x\}$ such that $xQ^iy$.

We observe that $T_1 \cup T_2 \in \mathcal{B}(A \setminus \{x, y\})$, since otherwise $D^*(0)$ is violated for the partition $A \setminus \{x, y\}, \{x\}, \{y\}$ of $A$ and the g-partition $T_1 \cup T_2, S \setminus \{i\}, \{i\}$ of $N$ (as $S \in \mathcal{B}_m(x)$ and $\{i\} \notin \mathcal{B}(\{y\})$ by (10.5)). We consider a $T_1 \cup T_2$-profile $P^{T_1 \cup T_2}$ such that $t_1(P^i) = x$ and $t_2(P^i) = y$ for all $i \in T_1 \cup T_2$. Let $P^N = (P^{T_1 \cup T_2}, Q^S)$. Since $F(P^N) \in C(E^F, P^N)$, we have $F(P^N) \in \{x, y\}$. Moreover, $A \setminus \{x\} \notin E(S \setminus \{i\})$ since $S \in \mathcal{B}_m(x)$, and by maximality of $E$ this implies $\{x\} \in E(T_1 \cup T_2 \cup \{i\})$. So $y$ is dominated by $\{x\}$ via the coalition $T_1 \cup T_2 \cup \{i\}$ and therefore $F(P^N) = x$. Thus, the coalition $T_1 \cup T_2$ improves on $Q^N$ by the deviation $P^{T_1 \cup T_2}$, a contradiction.

*Case 2.* $t_m(Q^i) = x$ for all $i \in S$.

We now observe that $S^{(1)} \cup T_1 \in \mathcal{B}(C_1)$, since otherwise $D^*(0)$ is violated for the partition $C_1, \{x\}, C_2$ of $A$ and the g-partition $S^{(1)} \cup T_1, S^{(2)}, T_2$ of

$N$. Consider now an $S^{(1)} \cup T_1$-profile $P^{S^{(1)} \cup T_1}$ such that $C_2 \, P^i \, \{x\} \, P^i \, C_1$ for all $i \in S^{(1)} \cup T_1$. Let $P^N = (P^{S^{(1)} \cup T_1}, Q^{S^{(2)} \cup T_2})$. Clearly, $F(P^N) \notin C_1$. Also, since $A \setminus C_2 \notin E(T_2)$, we have $C_2 \in E(S \cup T_1)$ by maximality. Thus, $x$ is dominated by $C_2$ via the coalition $S \cup T_1$. Hence $F(P^N) \in C_2$. Thus, since $F(R^N) \in C_1$, the coalition $S^{(1)} \cup T_1$ improves on $Q^N$ by the deviation $P^{S^{(1)} \cup T_1}$, a contradiction. □

We are now able to prove the main result of this section.

**Theorem 10.4.5.** *Let the effectivity function $E$ have an exactly and strongly consistent representation and satisfy* (10.5). *Then $E$ satisfies the conditions* $D^*(0), D^*(1), \ldots, D^*(m-2)$.

*Proof.* In view of Lemmas 10.4.2 and 10.4.4 it is sufficient to prove the following claim:

*Claim.* For $2 \leq k \leq m-2$, if $E$ satisfies $D^*(k-1)$, then $E$ satisfies $D^*(k)$.

To show this, suppose that $E$ satisfies $D^*(k-1)$ and suppose on the contrary that $D^*(k)$ is violated for the partition $\{x_1\}, \ldots, \{x_k\}, C_1, C_2$ of $A$ and the g-partition $S_1, \ldots, S_k, T_1, T_2$ of $N$. Since $S_i \in \mathcal{B}(x_i)$ and thus $A \setminus \{x_i\} \in E(S_i)$ we have $S_i \neq \emptyset$ for $i = 1, \ldots, k$, so that in particular $S_k \cup T_1 \neq N$. By superadditivity of $E$ we have (i) $S_k \cup T_1 \notin \mathcal{B}(\{x_k\} \cup C_1)$ or (ii) $N \setminus (S_k \cup T_1) \notin \mathcal{B}(A \setminus (\{x_k\} \cup C_1)$. If (i) holds then $D^*(k-1)$ is violated for the partition $\{x_1\}, \ldots, \{x_{k-1}\}, \{x_k\} \cup C_1, C_2$ of $A$ and the g-partition $S_1, \ldots, S_{k-1}, S_k \cup T_1, T_2$ of $N$. If (ii) holds, then $D^*(1)$ is violated for the partition $\{x_k\}, C_1, A \setminus (\{x_k\} \cup C_1)$ of $A$ and the g-partition $S_k, T_1, N \setminus (S_k \cup T_1)$ of $N$. □

We conclude this section with an example.

*Example 10.4.6.* Let $N = \{1, 2, 3, 4\}$ and $A = \{a, b, c\}$. Define an anonymous and neutral effectivity function $E$ by $E(\emptyset) = \emptyset$; $E(S) = \{A\}$ if $|S| = 1$; $E(S) = \{B \subseteq A \mid |B| \geq 2\}$ if $|S| = 2$; and $E(S) = P_0(A)$ if $|S| \geq 3$. Consider the partitions $\{1, 2\}, \{3\}, \{4\}$ and $\{a\}, \{b\}, \{c\}$. Then $\{1, 2\} \in \mathcal{B}_m(a)$, $\{3\} \notin \mathcal{B}(b)$, and $\{4\} \notin \mathcal{B}(c)$. Thus, $E$ violates $D^*(1)$. Since $E$ satisfies (10.5), it has no representation by an ESC social choice function.

## 10.5 Necessity of elimination stability for the existence of ESC representations

In this section we first show that under the (for our purposes) relevant assumptions on an effectivity function $E$ the conditions $D^*(k)$ of Section 10.4 imply the conditions $D(k)$ of Section 10.3.

**Lemma 10.5.1.** *Let the effectivity function $E$ be stable and maximal and satisfy the conditions $D^*(0), \ldots, D^*(m-2)$. Then $E$ satisfies $D(0), \ldots, D(m-2)$.*

*Proof.* Clearly, $D^*(0)$ implies $D(0)$. We shall prove that $D^*(k)$ implies $D(k)$ for $k = 1, \ldots, m-2$. Suppose, on the contrary, that $D(k)$ is violated by the enumeration $x_1, \ldots, x_k, x_{k+1}, \ldots, x_m$ of $A$ and the g-partition $S_1, \ldots, S_k, S_{k+1}, \ldots, S_m$ of $N$. Then $A \setminus \{x_{k+1}\} \notin E(S_{k+1})$ and $A \setminus \{x_{k+2}\} \notin E(S_{k+2})$.[1] By maximality of $E$, $\{x_{k+1}\} \in E(N \setminus S_{k+1})$ and $\{x_{k+2}\} \in E(N \setminus S_{k+2})$. Now $E$ is convex since it is maximal and stable (see Definition 5.3.3). Hence

$$\{x_{k+1}, x_{k+2}\} \in E(N \setminus (S_{k+1} \cup S_{k+2})). \tag{10.10}$$

Thus, by superadditivity of $E$ (which is implied by convexity), $A \setminus \{x_{k+1}, x_{k+2}\} \notin E(S_{k+1} \cup S_{k+2})$, so $S_{k+1} \cup S_{k+2} \notin \mathcal{B}(\{x_{k+1}, x_{k+2}\})$. If $k = m-3$, then we obtain a violation of $D^*(k)$, namely by taking $T_1 = S_{k+1} \cup S_{k+2}$ and $T_2 = S_{k+3}$ in Definition 10.4.3. Otherwise, by repeating the above argument we can show that $S_{k+1} \cup S_{k+2} \cup S_{k+3} \notin \mathcal{B}(\{x_{k+1}, x_{k+2}, x_{k+3}\})$ by deriving the associated version of (10.10). If $k = m-4$ we are done. Otherwise, we continue in the same manner until we obtain a violation of $D^*(k)$. □

The main result of this section is a direct consequence of Lemma 10.5.1 and Theorem 10.4.5.

**Theorem 10.5.2.** *Let the effectivity function $E$ satisfy* (10.5). *If $E$ has an exactly and strongly consistent representation then it satisfies $D(0), \ldots, D(m-2)$.*

The following corollary nicely summarizes some of the main findings of this chapter. It follows by combining Theorems 10.5.2 and 10.3.2, and Corollary 10.2.4.

**Corollary 10.5.3.** *Let the effectivity function $E$ be maximal and stable and satisfy* (10.5). *Then $E$ has an exactly and strongly consistent representation if and only if it is elimination stable.*

We conclude with an example which relates the results of this chapter to the results of Chapter 9 concerning anonymous effectivity functions.

*Example 10.5.4.* Let $E : P(N) \to P(P_0(A))$ be a monotonic and anonymous effectivity function. For $x \in A$ let

$$\beta(x) = \min\{|S| \mid A \setminus \{x\} \in E(S)\}.$$

Suppose that that $E$ has an ESC representation and that (10.5) holds or, equivalently, $\beta(x) \geq 2$ for all $x \in A$. We claim that $\sum_{x \in A} \beta(x) = n+1$. Clearly, $\sum_{x \in A} \beta(x) \geq n+1$ since otherwise it is straightforward to construct a preference profile $R^N \in L^N$ in which every alternative is blocked, so that $C(E, R^N) = \emptyset$, contradicting the stability of $E$. On the other hand, suppose that one of the conditions $D(k)$ is violated by the enumeration

[1] Note that for $k = m-2$ we obtain an immediate contradiction with $D^*(m-2)$.

$x_1, \ldots, x_k, x_{k+1}, \ldots, x_m$ and g-partition $S_1, \ldots, S_k, S_{k+1}, \ldots, S_m$. Then the number of players in $\cup_{j=k+1}^m S_j$ is equal to $n - \sum_{j=1}^k \beta(x_j)$ and this number must be smaller than or equal to $\sum_{j=k+1}^m (\beta(x_j) - 1)$, which is equivalent to $\sum_{x \in A} \beta(x) \geq n + m - k$. For none of the $D(k)$ to be violated (cf. Theorem 10.5.2) we therefore need that $\sum_{x \in A} \beta(x) < \min\{n + m - k \mid k = 0, \ldots, m-2\} = n+2$. Thus, $\sum_{x \in A} \beta(x) = n+1$. (Observe that $E_\beta(S) \subseteq E(S)$ for all $S \in P(N)$, where $E_\beta$ was defined in Section 9.3.)

Conversely, if $E$ is stable and maximal and $\sum_{x \in A} \beta(x) = n + 1$, then by the above argument $D(0), \ldots, D(m-2)$ are satisfied and therefore $E$ has an ESC representation by Theorem 10.3.2 and Corollary 10.2.4. Observe that (10.5) is not necessarily satisfied.

## 10.6 Notes and comments

This chapter is based mainly on Holzman (1986b). See also Ishikawa and Nakamura (1980). Holzman introduced the conditions $D(0), \ldots, D(m-2)$, which, if satisfied, guarantee the existence of an f.e.p. for every profile of preferences (linear orderings). His main contribution, however, is the investigation of the degree of necessity of these conditions: see Theorems 10.4.5 and 10.5.2.

We add that, if (10.5) does not hold, then the implication in Theorem 10.5.2 is not true. A simple counterexample is obtained by taking for $E$ a dictatorial effectivity function, i.e., there is $d \in N$ such that for all $S \in P_0(N)$, $E(S) = P_0(A)$ if $d \in S$ and $E(S) = \{A\}$ otherwise. Clearly, the dictatorial social choice function $F$ with dictator $d$ is an ESC representation of $E$, but $D(1)$ is violated (take $S_1 = \{d\}$ in Definition 10.3.1). Observe that (10.5) is violated as well. See also Holzman (1986b, p. 57) and Peleg (1978b), or Peleg (1984, Sect. 4.2).

Another important remark is that Example 10.5.4 shows that the analysis in Chapter 9 of anonymous effectivity functions is the most general possible. Finally, our writing of this chapter has benefitted from Abdou and Keiding (1991).

# Chapter 11
# Consistent voting systems with a continuum of voters

## 11.1 Motivation and summary

In this chapter we extend the model of Chapters 9 and 10 to a classical voting system with still finitely many alternatives (candidates) but with very many voters. Such a system is representative of political elections on the local or national level. As an, in our view, best approximation we model voters as elements of a non-atomic measure space. In particular, this approach allows us to accommodate the fact that in such voting systems single voters have negligible influence on the final outcome, and to avoid potential combinatorial complexities of a model with a large but finite number of voters.

The focus of the chapter is again on strategic aspects. If we talk about strategic aspects in this model, we necessarily deal with strategic voting by groups of voters (coalitions). This does not have to imply that voters in coalitions actually meet to coordinate their voting behavior. Although single voters are negligible for the final outcome, they may nevertheless derive utility from voting and, thus, may also vote strategically, possibly resulting in strategic behavior of groups of equally-minded voters.

After introducing the basics of the model in Section 11.2, we continue in Section 11.3 by showing that in this model the result of Gibbard (1973) and Satterthwaite (1975) persists. In particular, the requirement of non-manipulability implies the (undesirable) existence of an 'invisible dictator' as in Kirman and Sondermann (1972). Since, therefore, we cannot hope to reach the sincere outcome since we cannot expect voters to reveal their true preferences, we ask whether this outcome is at least attainable in an equilibrium of the voting game. Specifically, like in the preceding chapters we consider social choice functions satisfying the weaker requirement of exact and strong consistency (ESC). This means that for every given profile of preferences there is another profile which (i) is a strong (Nash) equilibrium – no coalition can profitably deviate – in the strategic game in which each voter

B. Peleg and H. Peters, *Strategic Social Choice*, Studies in Choice and Welfare, DOI 10.1007/978-3-642-13875-1_11, reports a preference and the outcome is evaluated according to given 'true' preferences, and (ii) results in the same alternative as the true preferences.

ESC social choice functions and associated effectivity functions are introduced in Section 11.4. We show that the main results of the model with finitely many voters go through: the effectivity function associated with an ESC social choice function is maximal, stable, and convex. This is no surprise: an ESC social choice function, seen as a game form, is a strong representation of the associated effectivity function, and the corresponding results of Chapter 5 continue to hold.

Next, we concentrate on anonymous ESC social choice functions, a natural restriction in large voting systems, and introduce blocking coefficients (Section 11.5) and feasible elimination procedures (Section 11.6). Here, our treatment deviates essentially from the case with finitely many voters. Sets of alternatives can be 'e-sets' or 'i-sets'. To block an i-set, a coalition needs to have size strictly larger than the blocking coefficient of that set, whereas for an e-set it can be larger or equal. Also, blocking coefficients constitute an additive function, contrary to the finitely many voters case (cf. Oren, 1981). As in Chapter 9, the main result is that any social choice function that selects maximal alternatives – that is, alternatives resulting from feasible elimination procedures – is exactly and strongly consistent. In Section 11.7 we establish equality of the core and the set of maximal alternatives for a collection of anonymous ESC social choice functions, and in Section 11.8 we show that this is actually a complete characterization of anonymous ESC social choice functions in case all blocking coefficients are required to be positive.

## 11.2 The basic model

Let $(\Omega, \Sigma, \lambda)$ be a non-atomic measure space. Here $\Omega$ is the set of *voters* or *players*; $\Sigma$ is the $\sigma$-field of permissible *coalitions*; and $\lambda$ is a nonnegative non-atomic measure on $\Sigma$, that is: $\lambda : \Sigma \to \mathbb{R}$ is a measure with $\lambda(S) \geq 0$ for all $S \in \Sigma$, and if $\lambda(S) > 0$ for some $S \in \Sigma$ then there is a $T \in \Sigma$ with $T \subseteq S$ and $0 < \lambda(T) < \lambda(S)$. The number $\lambda(S)$ for a coalition $S$ is interpreted as the size of $S$. By $\Sigma_0 = \Sigma \setminus \{\emptyset\}$ we denote the set of all nonempty coalitions, and by $\Sigma_+$ we denote the set of all coalitions $S$ with $\lambda(S) > 0$. Throughout we assume $\Omega \in \Sigma_+$ and $\lambda(\Omega) < \infty$.

Let $A$ be a finite set of *alternatives*. We assume throughout that $|A| \geq 3$. As before, a linear ordering of $A$ is a complete, reflexive, transitive, and antisymmetric binary relation on $A$, and the set of all linear orderings of $A$ is denoted by $L$.

A *profile* (of preferences) is a measurable function $\mathbf{R} : \Omega \to L$, that is, for each $R \in L$, $\{t \in \Omega \mid \mathbf{R}(t) = R\}$ is in $\Sigma$. Two profiles $\mathbf{R}_1$ and $\mathbf{R}_2$ are *equivalent*, written $\mathbf{R}_1 \sim \mathbf{R}_2$, if they differ only for a coalition of zero measure, that is: $\lambda(\{t \in \Omega \mid \mathbf{R}_1(t) \neq \mathbf{R}_2(t)\}) = 0$. Let $\rho$ denote the set of all profiles.

A *social choice function* (SCF) is a surjective function $F : \rho \to A$ that satisfies

$$\text{for all } \mathbf{R}_1, \mathbf{R}_2 \in \rho, \text{ if } \mathbf{R}_1 \sim \mathbf{R}_2, \text{ then } F(\mathbf{R}_1) = F(\mathbf{R}_2). \tag{11.1}$$

Condition (11.1) implies that social choice functions do not depend on the preferences of coalitions of measure 0. In particular, because of non-atomicity, single agents do not have any influence at all.

## 11.3 The Gibbard-Satterthwaite Theorem

In this section we show that the Gibbard-Satterthwaite Theorem continues to hold in our model with a continuum of voters, in the sense that any non-manipulable social choice function must exhibit a so-called *invisible dictator*. This is analogous to a similar result for Arrow's Impossibility Theorem in Kirman and Sondermann (1972). We start by formulating (non-)manipulability in the present context.

Let $\mathbf{R} \in \rho$ and $S \in \Sigma$. The social choice function $F$ is *manipulable by $S$ at $\mathbf{R}$* if there exists a $Q \in L$ with the following property: if $\mathbf{R}_1 \in \rho$ is a profile with $\mathbf{R}_1(t) = \mathbf{R}(t)$ for all $t \notin S$ and $\mathbf{R}_1(t) = Q$ for all $t \in S$, then $F(\mathbf{R}) \neq F(\mathbf{R}_1)$ and $F(\mathbf{R}_1)\mathbf{R}(t)F(\mathbf{R})$ for all $t \in S$.[1] Clearly, if $F$ is manipulable by $S$ at $\mathbf{R}$, then $\lambda(S) > 0$ by (11.1). We call $F$ *non-manipulable* if there exist no $\mathbf{R} \in \rho$ and $S \in \Sigma$ such that $F$ is manipulable by $S$ at $\mathbf{R}$. In words, it can never happen that all members of a coalition obtain a preferred alternative if that coalition coordinates on an untruthful preference. Observe that this non-manipulability condition has necessarily the form of coalitional non-manipulability since in our model single voters have no influence. Nevertheless, it can be weakened to a condition that is a closer approximation of individual non-manipulability. This is elaborated in Remark 11.3.7 below.

In order to formulate and prove the analogue of the Gibbard-Satterthwaite Theorem in this model we need to introduce the following concepts. A collection $\mathcal{D} \subseteq \Sigma_+$ is called an *ultrafilter* if (i) $D \cap D' \in \mathcal{D}$ for all $D, D' \in \mathcal{D}$ and (ii) $D \in \mathcal{D}$ or $\Omega \setminus D \in \mathcal{D}$ for every $D \in \Sigma_+$.[2] A *partition* of $\Omega$ is a finite collection of pairwise disjoint sets in $\Sigma_+$ the union of which has measure equal to $\lambda(\Omega)$.

Let $\mathcal{P} = \{D_1, \ldots, D_k\}$ be a partition of $\Omega$. Let $\mathcal{D}$ be an ultrafilter. We claim that there is at least one $i \in \{1, \ldots, k\}$ for which $D_i \in \mathcal{D}$. If not, then by property (ii) of $\mathcal{D}$, $D^i := \bigcup_{j=1,\ldots,k,\ j\neq i} D_j \in \mathcal{D}$ for every $i = 1, \ldots, k$, so by property (i), $\emptyset = \bigcap_{i=1,\ldots,k} D^i \in \mathcal{D}$, a contradiction since $\emptyset \notin \Sigma_+$. Hence,

[1] Requiring the voters in $S$ to coordinate on the same preference $Q$ in this definition is without loss of generality, as is not difficult to show.

[2] Observe that by (i) exactly one of the two statements in (ii) must hold.

there is an $i$ with $D_i \in \mathcal{D}$ and by property (i) again there is exactly one such $i$. Also, if a partition $\mathcal{P}'$ of $\Omega$ is coarser than $\mathcal{P}$ (i.e., each element of $\mathcal{P}$ is contained in an element of $\mathcal{P}'$; we also say that $\mathcal{P}$ is finer than $\mathcal{P}'$) then (i) implies $D \subseteq D'$, where $D$ and $D'$ are the elements of $\mathcal{P}$ and $\mathcal{P}'$ that are in $\mathcal{D}$, respectively. Therefore, there is a well defined mapping $d$ that assigns to each partition its element in $\mathcal{D}$, and $d$ satisfies:

$$\text{If } \mathcal{P}' \text{ is coarser than } \mathcal{P}, \text{ then } d(\mathcal{P}) \subseteq d(\mathcal{P}'). \tag{11.2}$$

The following lemma shows that also the converse holds.

**Lemma 11.3.1.** *Let $d$ be a mapping that assigns to each partition of $\Omega$ exactly one element of its elements. Suppose $d$ satisfies* (11.2). *Then the collection*

$$\mathcal{D} = \{D \in \Sigma_+ \mid \textit{there is a partition } \mathcal{P} \textit{ of } \Omega \textit{ with } D = d(\mathcal{P})\}$$

*is an ultrafilter.*

*Proof.* Let $\mathcal{P}^1$ and $\mathcal{P}^2$ be partitions and $D^1 = d(\mathcal{P}^1)$, $D^2 = d(\mathcal{P}^2)$. We show that $D^1 \cap D^2 \in \mathcal{D}$. Consider the join $\mathcal{P}$ of $\mathcal{P}^1$ and $\mathcal{P}^2$, i.e., the partition

$$\mathcal{P} = \{D \cap E \mid D \in \mathcal{P}^1,\ E \in \mathcal{P}^2,\ D \cap E \in \Sigma_+\}.$$

Obviously, $\mathcal{P}$ is finer than both $\mathcal{P}^1$ and $\mathcal{P}^2$. Suppose $D^* = d(\mathcal{P})$. Then by (11.2), both $D^* \subseteq D^1$ and $D^* \subseteq D^2$, hence $D^* \subseteq D^1 \cap D^2$. By definition of $\mathcal{P}$ therefore, $D^* = D^1 \cap D^2$, which implies $D^1 \cap D^2 \in \mathcal{D}$.

Finally let $D \in \Sigma_+$. If $\lambda(D) = \lambda(\Omega)$ then $D = d(\{D\})$, so $D \in \mathcal{D}$. Otherwise, either $D = d(\{D, \Omega \setminus D\})$ or $\Omega \setminus D = d(\{D, \Omega \setminus D\})$, hence either $D \in \mathcal{D}$ or $\Omega \setminus D \in \mathcal{D}$.

Thus, $\mathcal{D}$ is an ultrafilter. □

Now let $R_1, \ldots, R_{|A|!}$ be an enumeration of the elements of $L$. Each profile $\mathbf{R} \in \rho$ results in a collection $\mathcal{P} = \{S_1, \ldots, S_{|A|!}\}$ of subsets of of $\Omega$ with $S_k = \{t \in \Omega \mid \mathbf{R}(t) = R_k\} \in \Sigma$ for each $1 \leq k \leq |A|!$ . We denote by $\mathcal{P}(\mathbf{R})$ the collection obtained from $\mathcal{P}$ by omitting the sets of measure 0 and call this the *partition generated by* $\mathbf{R}$.

We associate with an ultrafilter $\mathcal{D}$ a social choice function $F^{\mathcal{D}}$, as follows. For a profile $\mathbf{R} \in \rho$ let $D$ be the unique element of $\mathcal{P}(\mathbf{R})$ that is in $\mathcal{D}$. Define $F^{\mathcal{D}}(\mathbf{R}) := x$ where $x\,R\,y$ for all $y \in A$ and $R = \mathbf{R}(t)$ for (all) $t \in D$. We have:

**Lemma 11.3.2.** *Let $\mathcal{D}$ be an ultrafilter. Then the social choice function $F^{\mathcal{D}}$ is non-manipulable.*

*Proof.* Let $\mathbf{R} \in \rho$. Suppose that coalition $S$ can manipulate at $\mathbf{R}$. Then $S \cap D = \emptyset$, where $D$ is the element of $\mathcal{P}(\mathbf{R})$ in $\mathcal{D}$. Hence, a manipulation of $S$ results in a profile $\mathbf{R}'$ such that $\mathcal{P}(\mathbf{R}')$ shares $D$ with $\mathcal{P}(\mathbf{R})$. But then $D$ is also the element of $\mathcal{P}(\mathbf{R}')$ that is in $\mathcal{D}$ by condition (i) of an ultrafilter. So $F^{\mathcal{D}}(\mathbf{R}') = F^{\mathcal{D}}(\mathbf{R})$, a contradiction. □

Conversely, let $F$ be a non-manipulable social choice function. We will show that there is an ultrafilter $\mathcal{D}$ such that $F = F^{\mathcal{D}}$, by applying (the Gibbard-Satterthwaite) Theorem 8.2.1. In order to satisfy the range condition in the theorem, we fix profiles $\mathbf{R}_1, \ldots, \mathbf{R}_{|A|}$ in $\rho$ such that $|\{F(\mathbf{R}_j) \mid j = 1, \ldots, |A|\}| = |A|$ – this is possible since $F$ is surjective by assumption. For an arbitrary partition $\mathcal{P} \subseteq \Sigma_+$ of $\Omega$ let $\mathcal{P}^*$ be the coarsest common refinement of $\mathcal{P}$ and the generated partitions $\mathcal{P}(\mathbf{R}_j)$, $j = 1, \ldots, |A|$. Regard every element of $\mathcal{P}^*$ as a separate agent. By Theorem 8.2.1 there is a fixed element $D^*$ of $\mathcal{P}^*$ such that, for every profile $\mathbf{R} \in \rho$ that is measurable with respect to $\mathcal{P}^*$, we have $F(\mathbf{R}) = x$ where $x$ is the top element of $\mathbf{R}(t)$ for (all) $t \in D^*$. Denote by $d^F(\mathcal{P})$ the element of $\mathcal{P}$ that contains $D^*$ and let

$$\mathcal{D}^F := \{d^F(\mathcal{P}) \mid \mathcal{P} \subseteq \Sigma_+ \text{ is a partition}\}.$$

**Lemma 11.3.3.** *(i) $\mathcal{D}^F$ is an ultrafilter. (ii) $F = F^{\mathcal{D}^F}$.*

*Proof.* (i) By Lemma 11.3.1 it is sufficient to prove that $d^F$ satisfies (11.2). Let $\mathcal{P}$ and $\mathcal{P}'$ be partitions with $\mathcal{P}'$ coarser than $\mathcal{P}$. Let $D' \in \mathcal{P}'$ with $d^F(\mathcal{P}) \subseteq D'$. Let $R, Q \in L$ have different top elements. Take a profile $\mathbf{R} \in \rho$ that is measurable with respect to $\mathcal{P}'$, and hence with respect to $\mathcal{P}$, and with $\mathbf{R}(t) = R$ for all $t \in D'$ and with $\mathbf{R}(t) = Q$ otherwise. Then $F(\mathbf{R})$ is the top element of $R$ since $R = \mathbf{R}(t)$ for (all) $t \in d^F(\mathcal{P})$. Hence, $d^F(\mathcal{P}') = D'$, so that $d^F(\mathcal{P}) \subseteq d^F(\mathcal{P}')$.

(ii) Let $\mathbf{R} \in \rho$ with generated partition $\mathcal{P}(\mathbf{R})$. Let $D^*$ be the element of $\mathcal{P}(\mathbf{R})^*$ such that $F(\mathbf{R}) = x$, where $x$ is the top element of $\mathbf{R}(t)$ for (all) $t \in D^*$. Let $D$ be the element of $\mathcal{P}(\mathbf{R})$ with $D^* \subseteq D$. By definition, $F^{\mathcal{D}^F}(\mathbf{R})$ is the top element of $\mathbf{R}(t)$ for (all) $t \in D$, hence $F^{\mathcal{D}^F}(\mathbf{R}) = x = F(\mathbf{R})$. □

Lemmas 11.3.2 and 11.3.3 have the following corollary.

**Corollary 11.3.4.** *Let $F : \rho \to A$ be a social choice function. Then $F$ is non-manipulable if and only if there is an ultrafilter $\mathcal{D}$ with $F = F^{\mathcal{D}}$.*

Corollary 11.3.4 is the form the Gibbard-Satterthwaite Theorem takes in our model with a continuum of voters and measurable profiles.[3] First, we show that the result is not vacuous.

**Theorem 11.3.5.** *There exists a non-manipulable social choice function.*

*Proof.* By Corollary 11.3.4 it is sufficient to show that there exists an ultrafilter of sets in $\Sigma_+$.

A *filter* in $\Sigma_+$ is a collection $\mathcal{F} \subseteq \Sigma_+$ satisfying

(i) for all $D, D' \in \mathcal{F}$, $D \cap D' \in \mathcal{F}$;

[3] For the case of finitely many voters the relation between the concepts of non-manipulability and ultrafilter has been examined before, see Batteau, Blin, and Monjardet (1981).

(ii) for all $D \in \mathcal{F}$ and $D' \in \Sigma_+$ with $D \subseteq D'$, $D' \in \mathcal{F}$.

(Clearly, an ultrafilter is a filter.) Let $\mathcal{U}$ be the collection of all filters $\mathcal{F}$ that satisfy, additionally,

(iii) for all $D \in \mathcal{F}$ and $D' \in \Sigma_+$ with $D' \subseteq D$ and $\lambda(D) = \lambda(D')$, $D' \in \mathcal{F}$.

Any set of positive measure together with all its subsets of the same measure and all measurable supersets of these form a filter, so $\mathcal{U}$ is non-empty. The inclusion relation is a partial ordering on $\mathcal{U}$ and each chain in $\mathcal{U}$ has an upper bound, namely the union of all filters in the chain. Hence, Zorn's Lemma implies that $\mathcal{U}$ has a maximal element, say $\mathcal{D}$. We claim that $\mathcal{D}$ is an ultrafilter. If not, then there is a $D \in \Sigma_+$ such that $D \notin \mathcal{D}$ and $\Omega \setminus D \notin \mathcal{D}$ (recall that $D \in \mathcal{D}$ and $\Omega \setminus D \in \mathcal{D}$ is not possible by (i)). By (ii), we have $D' \cap D \neq \emptyset$ and $D' \cap (\Omega \setminus D) \neq \emptyset$ for every $D' \in \mathcal{D}$ and by (iii), we have $\lambda(D' \cap D) > 0$ and $\lambda(D' \cap (\Omega \setminus D)) > 0$. Now consider the collection $\mathcal{D}'$ obtained by adding to $\mathcal{D}$ the collection $\{D' \cap D \mid D' \in \mathcal{D}\}$. Then it is easy to check that $\mathcal{D}'$ is a filter in $\mathcal{U}$ that is larger than $\mathcal{D}$, contradicting the maximality of $\mathcal{D}$. Hence, $\mathcal{D}$ is an ultrafilter. □

Since this existence proof is based on an application of Zorn's Lemma, it does not actually show how to construct a non-manipulable social choice function. If we require constructibility then it can be shown that a non-manipulable social choice function does not exist, so that Corollary 11.3.4 is truly an impossibility result. Observe that in our model a single voter cannot be a dictator in view of (11.1).

For a concrete illustration of Theorem 11.3.5 see the next example.

*Example 11.3.6.* Let $\Omega = [0, 1]$ and let $\lambda$ be the Lebesgue measure. If $\mathcal{D}$ is an ultrafilter, then for any $t \in [0, 1]$ exactly one of the two intervals $[0, t]$ and $[t, 1]$ must be in $\mathcal{D}$. Suppose, for the sake of the argument, that this is always the lower one, $[0, t]$. Then for every positive $\varepsilon$, every element of $\mathcal{D}$ has an intersection of positive measure with $[0, \varepsilon]$. The point 0 is an *invisible dictator* in the sense of Kirman and Sondermann (1972). Of course, the singleton 0 does not have any power at all, but always needs, roughly, a coalition of positive measure in any arbitrarily small neighborhood to exercise its 'dictatorship'. In this sense, the social choice function $F^{\mathcal{D}}$ associated with $\mathcal{D}$ has an invisible dictator, namely voter 0.

We conclude this section by discussing a possible weakening of the non-manipulability condition.

*Remark 11.3.7.* Our non-manipulability condition can be weakened to a version that is a closer approximation of individual non-manipulability. Call $F$ *$\varepsilon$-manipulable* if for every $\varepsilon > 0$ there is a profile $\mathbf{R} \in \rho$ and a coalition $S \in \Sigma$ with $\lambda(S) < \varepsilon$ such that F is manipulable by $S$ at $\mathbf{R}$. Call $F$ *non-$\varepsilon$-manipulable* if it is not $\varepsilon$-manipulable. This means that there is an $\varepsilon > 0$ such that at no profile coalitions with size smaller than $\varepsilon$ can manipulate. Clearly, non-$\varepsilon$-manipulability is weaker than non-manipulability, hence for every ultrafilter $\mathcal{D}$ the social choice function $F^{\mathcal{D}}$ satisfies it. Conversely, suppose that $F$ is non-$\varepsilon$-manipulable. Take $\varepsilon > 0$ so small that no coalition of size smaller than $\varepsilon$ can ever manipulate, and take an arbitrary partition $\mathcal{P}_\varepsilon$ of $\Omega$ such that each element of $\mathcal{P}_\varepsilon$ has size smaller than $\varepsilon$. Modify the definition of $\mathcal{P}^*$ preceding Lemma 11.3.3 such that $\mathcal{P}^*$ is now the coarsest common refinement of $\mathcal{P}_\varepsilon$ and $\mathcal{P}(\mathbf{R}_j)$, $j = 1, \dots, |A|$. Then Lemma 11.3.3 and Corollary 11.3.4 continue to hold if we replace non-manipulability by non-$\varepsilon$-manipulability.

## 11.4 Exactly and strongly consistent social choice functions

In the preceding section we have seen that a version of the Gibbard-Satterthwaite Theorem continues to hold in our model with a continuum of voters. Like in Chapters 9 and 10, as an answer to this we shall study exactly and strongly consistent social choice functions. We start with defining this concept within the present model.

Let $F$ be a social choice function and observe that for every $\mathbf{R} \in \rho$ the pair $(F, \mathbf{R})$ defines a game in strategic form in the usual and natural way: each player $t \in \Omega$ has strategy set $L$ and preference $\mathbf{R}(t)$ on $A$ for evaluating any outcome $F(\mathbf{R}^*) \in A$, $\mathbf{R}^* \in \rho$. For $S \in \Sigma_0$, denote by $\rho^S$ the set of all measurable functions $\mathbf{R}^S : S \to L$. Let $\mathbf{R} \in \rho$. The profile $\mathbf{Q} \in \rho$ is a *strong (Nash) equilibrium* of the game $(F, \mathbf{R})$ if for every $S \in \Sigma_+$ and every $\mathbf{V}^S \in \rho^S$, there exists $T \in \Sigma_+$ with $T \subseteq S$ and $F(\mathbf{Q})\,\mathbf{R}(t)\,F(\mathbf{Q}^{\Omega \setminus S}, \mathbf{V}^S)$ for every $t \in T$.

**Definition 11.4.1.** The social choice function $F$ is *exactly and strongly consistent* (ESC) if for every $\mathbf{R} \in \rho$ there exists a strong equilibrium $\mathbf{Q}$ of $(F, \mathbf{R})$ such that $F(\mathbf{Q}) = F(\mathbf{R})$.

Thus, if $F$ is an ESC social choice function, then for every profile there is a strong equilibrium profile that results in the same outcome, and therefore $F$ is not necessarily distorted.

In the remainder of this chapter we shall concentrate on anonymous ESC social choice functions. In our model a social choice function $F : \rho \to A$ is *anonymous* if for all $\mathbf{R}_1, \mathbf{R}_2 \in \rho$ we have: if $\lambda(\{t \in \Omega \mid \mathbf{R}_1(t) = R\}) = \lambda(\{t \in \Omega \mid \mathbf{R}_2(t) = R\})$ for all $R \in L$, then $F(\mathbf{R}_1) = F(\mathbf{R}_2)$. Thus, a social choice function is anonymous if it only depends on the numbers of voters for each preference.

We first consider a simple example of an anonymous ESC social choice function.[4] For a profile $\mathbf{R}$ and $a, b \in A$, $a \neq b$, we say that *$a$ Pareto dominates $b$* if $\lambda(\{t \in \Omega \mid b\,\mathbf{R}(t)\,a\}) = 0$. We call an alternative $a \in A$ *Pareto optimal*

[4] This example is similar to Example 5.2.2.

with respect to $\mathbf{R}$ if it is not Pareto dominated by some other element of $A$, and denote by PAR($\mathbf{R}$) the set of Pareto optimal alternatives with respect to $\mathbf{R}$.

*Example 11.4.2.* Let $\bar{a} \in A$ be a designated alternative, and let $R_0 \in L$ be fixed. Define a social choice function $F : \rho \to A$ by

$$F(\mathbf{R}) = \begin{cases} \bar{a} \text{ if } \bar{a} \in \mathrm{PAR}(\mathbf{R}) \\ a \text{ if } \bar{a} \notin \mathrm{PAR}(\mathbf{R}) \text{ and } a \text{ is the } R_0\text{-maximum} \\ \qquad \text{of } \{b \in \mathrm{PAR}(\mathbf{R}) \mid b \text{ Pareto dominates } \bar{a}\} \end{cases}$$

for all $\mathbf{R} \in \rho$. Note that $\bar{a}$ can be interpreted as the 'status quo'. Obviously, $F$ is surjective and anonymous. We show that $F$ is ESC. Let $\mathbf{R} \in \rho$. We distinguish the following possibilities.

(i) $\bar{a} \in \mathrm{PAR}(\mathbf{R})$.
Let $\mathbf{Q} \in \rho$ satisfy $\bar{a}\,\mathbf{Q}(t)\,a$ for all $t \in \Omega$ and $a \in A \setminus \{\bar{a}\}$. Then $\mathbf{Q}$ is a strong equilibrium of $(F, \mathbf{R})$ and $F(\mathbf{Q}) = F(\mathbf{R})$.

(ii) $\bar{a} \notin \mathrm{PAR}(\mathbf{R})$.
Let $q$ be the $R_0$-maximum of $B = \{b \in \mathrm{PAR}(\mathbf{R}) \mid b \text{ Pareto dominates } \bar{a}\}$. Define $\mathbf{Q} \in \rho$ by $q\,\mathbf{Q}(t)\,\bar{a}\,\mathbf{Q}(t)\,a$ for all $t \in \Omega$ and $a \in A \setminus \{\bar{a}, q\}$. Then $F(\mathbf{Q}) = q = F(\mathbf{R})$ and $\mathbf{Q}$ is a strong equilibrium of $(F, \mathbf{R})$. Indeed, $\Omega$ does not have a profitable deviation from $\mathbf{Q}$ since $q$ is Pareto optimal with respect to $\mathbf{R}$. Now let $S \in \Sigma_+$, $\lambda(S) < \lambda(\Omega)$, and $\mathbf{V}^S \in \rho^S$. Then $F(\mathbf{Q}^{\Omega \setminus S}, \mathbf{V}^S) \in \{\bar{a}, q\}$. Hence, $\mathbf{V}^S$ cannot be a profitable deviation for $S$.

## 11.4.1 Effectivity functions of ESC social choice functions

Before proceeding with our investigation of anonymous ESC social choice functions, we define the concept of an effectivity function in our model, and collect some properties of effectivity functions associated with ESC social choice functions – analogous to the case with finitely many voters in Chapter 10.

**Definition 11.4.3.** An *effectivity function* (EF) is a function $E : \Sigma \to P(P_0(A))$ that satisfies the following conditions: (i) $E(\Omega) = P_0(A)$; (ii) $E(\emptyset) = \emptyset$; (iii) $A \in E(S)$ for every $S \in \Sigma_0$; and (iv) if $S_1, S_2 \in \Sigma_0$ and $\lambda(S_1 \setminus S_2) + \lambda(S_2 \setminus S_1) = 0$, then $E(S_1) = E(S_2)$.

Condition (iv) in Definition 11.4.3 is specific for our model. It says that the effectivity function does not distinguish between coalitions that differ only in a set of measure 0.

All of the following definitions and statements are analogous to their counterparts in the finite case, but we nevertheless list them for the sake of completeness.

An effectivity function $E$ is *superadditive* if for all $S_1, S_2 \in \Sigma$ with $S_1 \cap S_2 = \emptyset$ and all $B_1 \in E(S_1)$ and $B_2 \in E(S_2)$ we have: $B_1 \cap B_2 \in E(S_1 \cup S_2)$. The EF $E$ is *monotonic* if for all $S, S^* \in \Sigma$ and $B, B^* \in P_0(A)$ with $B \in E(S)$, $S \subseteq S^*$ and $B \subseteq B^*$, we have $B^* \in E(S^*)$. An EF $E$ is *maximal* if for all $S \in \Sigma_0$ and $B \in P_0(A)$ we have: if $B \notin E(S)$ then $A \setminus B \in E(\Omega \setminus S)$. An EF $E$ is *convex* if for all $S_1, S_2 \in \Sigma$ and $B_1 \in E(S_1)$, $B_2 \in E(S_2)$ we have $B_1 \cap B_2 \in E(S_1 \cup S_2)$ or $B_1 \cup B_2 \in E(S_1 \cap S_2)$.

Also the core of an effectivity function is defined exactly as in the finite model. Let $E : \Sigma \to P(P_0(A))$ be an EF and let $\mathbf{R} \in \rho$. Let $B \in P_0(A)$, $x \in A \setminus B$, and $S \in \Sigma$. We say that $B$ *dominates* $x$ *via* $S$ *at* $\mathbf{R}$ if $B \in E(S)$ and $b\,\mathbf{R}(t)\,x$ for all $b \in B$ and $t \in S$. Also, $x$ is *dominated at* $\mathbf{R}$ if there exists $B \in P_0(A)$ and $S \in \Sigma$ such that $B$ dominates $x$ via $S$ at $\mathbf{R}$. If $b$ is not dominated at $\mathbf{R}$ then $b$ is *undominated* at $\mathbf{R}$.

**Definition 11.4.4.** The *core* $C(E, \mathbf{R})$ is the set of all undominated alternatives at $\mathbf{R}$. The effectivity function $E$ is *stable* if $C(E, \mathbf{R}) \neq \emptyset$ for all $\mathbf{R} \in \rho$.

Let $F : \rho \to A$ be a social choice function. We associate with $F$ an effectivity function $E^F$ as follows. Let $S \in \Sigma_0$ and let $B \in P_0(A)$. Call $S$ effective for $B$ if there exists an $\mathbf{R}^S \in \rho^S$ such that $F(\mathbf{R}^S, \mathbf{Q}^{\Omega\setminus S})$ is in $B$ for every $\mathbf{Q}^{\Omega\setminus S} \in \rho^{\Omega\setminus S}$. Define $E^F(\emptyset) = \emptyset$, and for $S \in \Sigma \setminus \{\emptyset\}$

$$E^F(S) = \{B \in P_0(A) \mid S \text{ is effective for } B\}.$$

In the following theorem we collect some useful properties of $E^F$.

**Theorem 11.4.5.** *Let $F : \rho \to A$ be an ESC social choice function. Then $E^F$ is superadditive, monotonic, maximal, stable, and convex. Moreover, $F(\mathbf{R}) \in C(E^F, \mathbf{R})$ for all $\mathbf{R} \in \rho$.*

*Proof.* Superadditivity and monotonicity are straightforward from the definition of $E^F$ (ESC is not needed for this). Maximality and stability, as well as the last statement in the theorem, can be proved analogously to the case of finitely many voters, see Section 5.2. Finally, stability and maximality together imply convexity. A proof of this fact is analogous to the proof of Theorem 6.A.9 in Peleg (1984). □

## 11.5 Blocking coefficients of anonymous ESC SCFs

In the remainder of the chapter we concentrate on anonymous ESC social choice functions. Anonymity is a natural requirement for voting procedures. Moreover, imposing this condition will enable us to derive much more detailed results on both social choice functions and effectivity functions than Theorem 11.4.5 provides.

Let $F : \rho \to A$ be an anonymous ESC social choice function, with associated effectivity function $E^F$. Then $E^F$ is superadditive, monotonic, maximal, stable and convex, cf. Theorem 11.4.5. A central concept is that of a blocking coefficient.

For $B \in P_0(A) \setminus \{A\}$, the *blocking coefficient* is the real number

$$\beta(B) = \inf\{\lambda(S) \mid A \setminus B \in E^F(S)\}. \tag{11.3}$$

The number $\beta(B)$ represents the minimum size of a 'blocking coalition' of $B$. It is useful since $F$ is anonymous. We call $B$ an *e-set* ('e' from 'equality') if $S \in \Sigma$ and $\lambda(S) = \beta(B)$ imply that $A \setminus B \in E^F(S)$; otherwise, $B$ is called an *i-set* ('i' from 'inequality'). Thus, to block an e-set $B$ we need a coalition of size at least $\beta(B)$ but to block an i-set $B$ we need a coalition of size strictly larger than $\beta(B)$.

We formulate a first observation concerning the blocking coefficients $\beta(\cdot)$. Suppose $B \in P_0(A)$ is an e-set. If $\beta(B) = 0$ then $A \setminus B \in E^F(S)$ for some coalition $S \in \Sigma_0$ with $\lambda(S) = 0$. Since $B \in E^F(\Omega \setminus S)$ by conditions (i) and (iv) in the definition of an effectivity function, we have a violation of superadditivity of $E^F$. Thus, we have shown:

$$\text{If } B \text{ is an e-set, then } \beta(B) > 0. \tag{11.4}$$

We now derive a number of other properties of $\beta(\cdot)$, in particular Theorem 11.5.1 below, which says that $\beta(\cdot)$ is an additive function.

If $B_1, B_2 \in P_0(A)$ and $B_1 \cup B_2 \neq A$, then

$$\beta(B_1 \cup B_2) \leq \beta(B_1) + \beta(B_2). \tag{11.5}$$

To see this, note that if the right hand side of this inequality is greater than or equal to $\lambda(\Omega)$, then the inequality holds. Now assume it is smaller. Let $\varepsilon > 0$ be small and let $S_i \in \Sigma$ with $\lambda(S_i) = \beta(B_i) + \varepsilon$ and $A \setminus B_i \in E^F(S_i)$ for $i = 1, 2$, such that $S_1 \cap S_2 = \emptyset$. By superadditivity, $A \setminus (B_1 \cup B_2) \in E^F(S_1 \cup S_2)$, hence $\beta(B_1 \cup B_2) \leq \beta(B_1) + \beta(B_2) + 2\varepsilon$. By letting $\varepsilon$ approach 0, (11.5) follows.

For every $B \in P_0(A) \setminus \{A\}$ we have

$$\beta(B) + \beta(A \setminus B) \geq \lambda(\Omega) \tag{11.6}$$

because otherwise there would be disjoint coalitions $S$ and $T$ with $B \in E^F(S)$ and $A \setminus B \in E^F(T)$, contradicting the superadditivity of $E^F$. We shall now show the reverse inequality. Assume $\beta(B) > 0$ otherwise there is nothing left to prove. For every $0 < \delta < \beta(B)$ and $S \in \Sigma$ with $\lambda(S) = \delta$ we have $A \setminus B \notin E^F(S)$. Hence by maximality of $E^F$, $B \in E^F(\Omega \setminus S)$, so $\beta(A \setminus B) \leq \lambda(\Omega) - \delta$. This implies the reverse inequality of (11.6), hence

$$\beta(B) + \beta(A \setminus B) = \lambda(\Omega) \tag{11.7}$$

for every $B \in P_0(A) \setminus \{A\}$.

Suppose that $B \in P_0(A) \setminus \{A\}$ is an e-set and let $S \in \Sigma$ such that $\beta(B) = \lambda(S)$ and $A \setminus B \in E^F(S)$. Then, by superadditivity, $B \notin E^F(\Omega \setminus S)$. Also, by (11.7), $\beta(A \setminus B) = \lambda(\Omega \setminus S)$, so that $A \setminus B$ is an i-set. Conversely, let $A \setminus B$ be an i-set and $S \in \Sigma$ with $\lambda(S) = \beta(A \setminus B)$. Then $B \notin E^F(S)$ so that, by maximality, $A \setminus B \in E^F(\Omega \setminus S)$. Since, by (11.7), $\beta(B) = \lambda(\Omega \setminus S)$, we have that $B$ is an e-set. Summarizing,

$$B \text{ is an e-set} \Leftrightarrow A \setminus B \text{ is an i-set} \tag{11.8}$$

for every $B \in P_0(A) \setminus \{A\}$.

Moreover, monotonicity of $E^F$ clearly implies monotonicity of the function $\beta(\cdot)$:

$$B_1 \subseteq B_2 \Rightarrow \beta(B_1) \leq \beta(B_2) \tag{11.9}$$

for all $B_1, B_2 \in P_0(A) \setminus \{A\}$.

We now show that blocking coefficients are actually additive, that is, $\beta(B_1 \cup B_2) = \beta(B_1) + \beta(B_2)$ for all $B_1, B_2 \in P_0(A)$ with $B_1 \cap B_2 = \emptyset$ and $B_1 \cup B_2 \neq A$.

**Theorem 11.5.1.** *$\beta(\cdot)$ is additive.*

*Proof.* Let $B_i \in P_0(A)$, $i = 1, 2$, with $B_1 \cap B_2 = \emptyset$ and $B_1 \cup B_2 \neq A$. In view of (11.5) it is sufficient to prove that $\beta(B_1 \cup B_2) \geq \beta(B_1) + \beta(B_2)$. By (11.9) we may assume $\beta(B_i) > 0$ for $i = 1, 2$. Let $S$ and $T$ satisfy $\lambda(S) < \beta(B_1)$, $\lambda(T) < \beta(B_2)$, and $S \cap T = \emptyset$. Then by (11.7) and (11.9), $B_1 \in E^F(\Omega \setminus S)$ and $B_2 \in E^F(\Omega \setminus T)$. By convexity of $E^F$, $B_1 \cup B_2 \in E^F(\Omega \setminus (S \cup T))$. Thus, by (11.7) and the definition of $\beta(\cdot)$,

$$\begin{aligned}\beta(B_1 \cup B_2) &= \lambda(\Omega) - \beta(A \setminus (B_1 \cup B_2)) \\ &\geq \lambda(\Omega) - \lambda(\Omega \setminus (S \cup T)) \\ &= \lambda(S) + \lambda(T).\end{aligned}$$

Since, by (11.7) and (11.9), $\beta(B_1) + \beta(B_2) \leq \lambda(\Omega)$, we can choose $\lambda(S)$ and $\lambda(T)$ as close to $\beta(B_1)$ and $\beta(B_2)$, respectively, as desired, which completes the proof. □

In view of Theorem 11.5.1 and (11.7) it is useful to define $\beta(A) = \lambda(\Omega)$ and let $A$ be an i-set. Note that this is another deviation from the case of finitely many voters, where the analogous statement is $\sum_{a \in A} \beta(a) = n + 1$, cf. Section 9.2.

For e-sets we have the following theorem.

**Theorem 11.5.2.** *If $B_1$ and $B_2$ are e-sets, then $B_1 \cap B_2$ or $B_1 \cup B_2$ are e-sets.*

*Proof.* Let $B_1$ and $B_2$ be e-sets. If $B_1 \cap B_2 = \emptyset$ then take disjoint coalitions $S_1$ and $S_2$ of sizes $\beta(B_1)$ and $\beta(B_2)$, respectively. Then $A \setminus B_1 \in E^F(S_1)$ and $A \setminus B_2 \in E^F(S_2)$. By superadditivity, $A \setminus (B_1 \cup B_2) \in E^F(S_1 \cup S_2)$. Since $\lambda(S_1 \cup S_2) = \beta(B_1 \cup B_2)$ by Theorem 11.5.1, we conclude that $B_1 \cup B_2$ is an e-set.

Next, assume $B_1 \cap B_2 \neq \emptyset$. Choose pairwise disjoint sets $S_1$, $S_2$, and $S_3$ in $\Sigma_0$ such that $\lambda(S_1) = \beta(B_1) - \beta(B_1 \cap B_2)$, $\lambda(S_2) = \beta(B_2) - \beta(B_1 \cap B_2)$, and $\lambda(S_3) = \beta(B_1 \cap B_2)$. Define $T_1 = S_1 \cup S_3$ and $T_2 = S_2 \cup S_3$. Then $\lambda(T_1) = \beta(B_1)$, $\lambda(T_2) = \beta(B_2)$, $\lambda(T_1 \cap T_2) = \beta(B_1 \cap B_2)$, and $\lambda(T_1 \cup T_2) = \beta(B_1 \cup B_2)$. By assumption, $A \setminus B_1 \in E^F(T_1)$ and $A \setminus B_2 \in E^F(T_2)$. Since $E^F$ is convex, $A \setminus (B_1 \cup B_2) \in E^F(T_1 \cup T_2)$ or $A \setminus (B_1 \cap B_2) \in E^F(T_1 \cap T_2)$. Thus, $B_1 \cup B_2$ or $B_1 \cap B_2$ are e-sets. □

*Example 11.5.3.* The effectivity function associated with the ESC social choice function of Example 11.4.2 is as follows. If $S \in \Sigma_+$ with $\lambda(S) < \lambda(\Omega)$ then $B \in E^F(S)$ if and only if $\bar{a} \in B$, for all $B \in P_0(A)$; and if $\lambda(S) = \lambda(\Omega)$ then $E^F(S) = P_0(A)$. This implies that for all $B \in P_0(A)$ we have $\beta(B) = 0$ if $\bar{a} \notin B$, and $\beta(B) = \lambda(\Omega)$ if $\bar{a} \in B$. Also, $B \neq A$ is an i-set if $\bar{a} \notin B$, and an e-set if $\bar{a} \in B$. In particular, $\beta(\{x\}) = 0$ and $\{x\}$ is an i-set for all $x \in A \setminus \{\bar{a}\}$, and $\beta(\{\bar{a}\}) = \lambda(\Omega)$ and $\{\bar{a}\}$ is an e-set.

We conclude this section by generalizing the concepts of e-sets and i-sets. Let $\beta : P_0(A) \to [0, \lambda(\Omega)]$ and let $\{\mathbf{i}, \mathbf{e}\}$ be a partition of $P_0(A)$ satisfying

$$\beta \text{ is additive, } \beta(A) = \lambda(\Omega), \text{ and } \beta(B) > 0 \text{ for all } B \in \mathbf{e}, \tag{11.10}$$

$$\text{for all } B \in P_0(A) \setminus \{A\},\ B \in \mathbf{e} \Leftrightarrow A \setminus B \in \mathbf{i}, \text{ and } A \in \mathbf{i}, \tag{11.11}$$

$$\text{for all } B_1, B_2 \in \mathbf{e}, \text{ we have } B_1 \cap B_2 \in \mathbf{e} \text{ or } B_1 \cup B_2 \in \mathbf{e}. \tag{11.12}$$

Properties (11.10)–(11.12) summarize exactly all the properties of the e-sets and i-sets of the effectivity function associated with an anonymous ESC social choice function established above.

Next, for a system $(\beta; \mathbf{e}, \mathbf{i})$ satisfying (11.10)–(11.12), we define an effectivity function $E$ by $E(\Omega) = P_0(A)$, $E(\emptyset) = \emptyset$, $A \in E(S)$ for every $S \in \Sigma_0$, and

$$\text{for all } B \in \mathbf{e} \text{ and } S \in \Sigma, \text{ if } \lambda(S) \geq \beta(B) \text{ then } A \setminus B \in E(S), \tag{11.13}$$

$$\text{for all } B \in \mathbf{i} \text{ and } S \in \Sigma, \text{ if } \lambda(S) > \beta(B) \text{ then } A \setminus B \in E(S). \tag{11.14}$$

It is straightforward to check that $E$ is an effectivity function according to Definition 11.4.3: the premise in condition (iv) implies in particular that $\lambda(S_1) = \lambda(S_2)$, so that $E(S_1) = E(S_2)$ according to the definition of $E$.

In the next sections we consider the following question. Given a system $(\beta; \mathbf{e}, \mathbf{i})$ satisfying (11.10)–(11.12) and associated effectivity function $E$, is there an (anonymous) ESC social choice function $F$ such that $E = E^F$? By using feasible elimination procedures we will present a complete answer to this question for the case where there is exactly one i-alternative, i.e., there is exactly one $x \in A$ with $\{x\} \in \mathbf{i}$, in Corollary 11.7.3. This is restrictive since we already know that there are other cases: see Examples 11.4.2, 11.5.3. On the other hand, this case is the only possible one if we require all blocking coefficients to be positive: see Corollary 11.8.3.

## 11.6 Feasible elimination procedures

In this section we describe a procedure that will enable the construction of an anonymous exactly and strongly consistent social choice function. We start with the definition of a so-called *pseudo* feasible elimination procedure.

Throughout, $\beta : A \to \mathbb{R}$ is a function satisfying $\beta(a) \geq 0$ for all $a \in A$ and $\sum_{a \in A} \beta(a) = \lambda(\Omega)$.

**Definition 11.6.1.** Let $\mathbf{R} \in \rho$. A *pseudo feasible elimination procedure* (p.f.e.p.) is a sequence $(x_1, C_1; \ldots; x_{m-1}, C_{m-1}; x_m)$ that satisfies the following conditions:

$$A = \{x_1, \ldots, x_m\}. \tag{11.15}$$

$$C_j \in \Sigma_0,\ C_j \cap C_k = \emptyset,\ \lambda(C_j) \geq \beta(x_j) \\ \text{for all } j, k = 1, \ldots, m-1,\ j \neq k. \tag{11.16}$$

$$y\,\mathbf{R}(t)\,x_j \text{ for all } j = 1, \ldots, m-1,\ y \in \{x_{j+1}, \ldots, x_m\},\ t \in C_j. \tag{11.17}$$

In a pseudo feasible elimination procedure, 'bottom' alternatives are eliminated consecutively. As $\sum_{a \in A} \beta(a) = \lambda(\Omega)$, it is obvious that for each profile there always exists at least one p.f.e.p., namely one with $\lambda(C_j) = \beta(x_j)$ for all $j = 1, \ldots, m-1$. In the following lemma we show that actually more is possible: if an alternative $x$ is bottom for a coalition of size larger than $\beta(x)$, then there is a p.f.e.p. where this alternative is eliminated first and with strict inequality.

First we recall a notation: for a profile $\mathbf{R}$ and a subset $B$ of $A$, denote by $\mathbf{R}_{|B}$ the profile of preferences restricted to the set $B$.

**Lemma 11.6.2.** *Let* $\mathbf{R} \in \rho$ *and let* $x \in A$ *satisfy*

$$\lambda(\{t \in \Omega \mid y\,\mathbf{R}(t)\,x \text{ for all } y \in A\}) > \beta(x).$$

*Then there exists a p.f.e.p.* $(x, C_x; x_1, C_1; \ldots; x_{m-1})$ *with* $\lambda(C_x) > \beta(x)$.

*Proof.* The proof is by induction on $m$. The case $m = 2$ is obvious. Let $m \geq 3$. We define

$$A^* = \{y \in A \mid \lambda(\{t \in \Omega \mid z\,\mathbf{R}(t)\,y \text{ for all } z \in A\}) > \beta(y)\}. \tag{11.18}$$

By assumption, $x \in A^*$. We distinguish the following cases.

(i) $|A^*| \geq 2$.

Let $y \in A^* \setminus \{x\}$ and choose $C_y \subseteq \Omega$ such that $\lambda(C_y) = \beta(y)$ and $C_y \subseteq \{t \in \Omega \mid z\,\mathbf{R}(t)\,y$ for all $z \in A\}$. Define the profile $\mathbf{Q} \in \rho$ as follows. If $t \in \Omega \setminus C_y$ with $z\,\mathbf{R}(t)\,y$ for all $z \in A$, then let $x\,\mathbf{Q}(t)\,A \setminus \{x, y\}\,\mathbf{Q}(t)\,y$; otherwise, $\mathbf{Q}(t) = \mathbf{R}(t)$. Consider the restricted profile $\mathbf{Q}_1 = \mathbf{Q}^{\Omega \setminus C_y}{}_{|A \setminus \{y\}}$ – observe that if $x$ is a bottom alternative for a voter $t$ in this restricted profile then it was a bottom element of $\mathbf{R}(t)$. By the induction hypothesis and by the construction of $\mathbf{Q}$ there exists a p.f.e.p. $(x, C_x; x_1, C_1; \ldots; x_{m-2})$ with respect to $\mathbf{Q}_1$ such that $\lambda(C_x) > \beta(x)$ and $C_x \subseteq \{t \in \Omega \mid z\,\mathbf{R}(t)\,x$ for all $z \in A\}$. Then the p.f.e.p. $(x, C_x; y, C_y; x_1, C_1; \ldots; x_{m-2})$ is as required.

(ii) $A^* = \{x\}$.

Let $\hat{C}_x$ satisfy $\hat{C}_x \subseteq \{t \in \Omega \mid y\,\mathbf{R}(t)\,x$ for all $y \in A\}$ and $\lambda(\hat{C}_x) = \beta(x)$. Consider the profile $\mathbf{R}_1 = \mathbf{R}^{\Omega\setminus\hat{C}_x}|_{A\setminus\{x\}}$. For all $y \neq x$ let $C_y = \{t \in \Omega \setminus \hat{C}_x \mid z\,\mathbf{R}(t)\,y$ for all $z \in A \setminus \{x\}\}$. We distinguish two subcases.

(ii.1) $\lambda(C_y) = \beta(y)$ for all $y \neq x$.
Choose $\bar{y} \in A \setminus \{x\}$ such that $\lambda(\hat{C}) > 0$, where

$$\hat{C} = \{t \in \Omega \setminus \hat{C}_x \mid z\,\mathbf{R}(t)\,\bar{y}\mathbf{R}(t)x \text{ for all } z \in A \setminus \{x\}\}.$$

(Observe that $\hat{C} \subseteq C_{\bar{y}}$, hence $\lambda(\hat{C}) \leq \beta(\bar{y})$.) Let $C_x = \hat{C}_x \cup \hat{C}$, and let $A \setminus \{x, \bar{y}\} = \{y_1, \ldots, y_{m-2}\}$. Then $(x, C_x; y_1, C_{y_1}; \ldots; y_{m-2}, C_{y_{m-2}}; \bar{y})$ is a p.f.e.p. as required.

(ii.2) There exists $\bar{y} \neq x$ such that $\lambda(C_{\bar{y}}) > \beta(\bar{y})$.
By the induction hypothesis there exists a p.f.e.p. $(\bar{y}, \hat{C}_{\bar{y}}; x_1, C_1; \ldots, x_{m-2})$ with respect to $\mathbf{R}_1$ such that $\lambda(\hat{C}_{\bar{y}}) > \beta(\bar{y})$. Note that $\lambda(\{t \in \hat{C}_{\bar{y}} \mid z\,\mathbf{R}(t)\,\bar{y}\,\mathbf{R}(t)\,x$ for all $z \in A \setminus \{x\}\}) > 0$ since $\bar{y} \notin A^*$. Choose $\hat{C} \subseteq \{t \in \hat{C}_{\bar{y}} \mid z\,\mathbf{R}(t)\,\bar{y}\,\mathbf{R}(t)\,x$ for all $z \in A \setminus \{x\}\}$ such that $0 < \lambda(\hat{C}) \leq \lambda(\hat{C}_{\bar{y}}) - \beta(\bar{y})$. Then $(x, \hat{C}_x \cup \hat{C}; \bar{y}, \hat{C}_{\bar{y}} \setminus \hat{C};\, x_1, C_1; \ldots; x_{m-2})$ is a p.f.e.p. as required. □

Now note that if a procedure like p.f.e.p. should result in an anonymous ESC social choice function then clearly some of the alternatives might be i-alternatives and these should be blocked with inequality. The preceding lemma exhibits a case in which this is possible. If, however, there are two or more of such i-alternatives then it is not difficult to construct a profile where a p.f.e.p. does not exist if we require i-alternatives to be blocked with inequality. With this consideration and with observation (11.4) – which says that only i-alternatives can have zero blocking coefficients – in mind, all alternatives except at most one should have positive $\beta$-values. Therefore, in the rest of this section we make the following assumption.

**Assumption 11.6.3** There is an alternative in $A$, denoted by $s$, such that $\beta(a) > 0$ for all $a \in A \setminus \{s\}$.

We next introduce the concept of a feasible elimination procedure within the model of this chapter. In this procedure, the designated alternative $s$ of Assumption 11.6.3 can only be eliminated if, during the procedure, it becomes a bottom alternative for a coalition of size strictly larger than $\beta(s)$.

**Definition 11.6.4.** Let $\mathbf{R} \in \rho$. A p.f.e.p. $(x_1, C_1; \ldots; x_{m-1}, C_{m-1}; x_m)$ is a *feasible elimination procedure* (f.e.p.) if it satisfies the following condition:

$$x_m = s \text{ or } [x_j = s \text{ for some } j < m \text{ and } \lambda(C_j) > \beta(s)]. \tag{11.19}$$

We shall now prove the existence of f.e.p.'s in our model and then relate them to ESC social choice functions.

**Theorem 11.6.5.** *Let Assumption 11.6.3 hold. Then, for every* $\mathbf{R} \in \rho$ *there is an f.e.p. with respect to* $\mathbf{R}$.

*Proof.* Let $\mathbf{R} \in \rho$. The proof is by induction on $m$. The case $m = 2$ is obvious. Let $m \geq 3$. Define $A^*$ as in (11.18). We distinguish the following possibilities.

(i) $A^* = \emptyset$.
For $a \in A$ let $C(a) = \{t \in \Omega \mid y\,\mathbf{R}(t)\,a$ for all $y \in A\}$. Then $\lambda(C(a)) = \beta(a)$ for all $a \in A$. Let $A \setminus \{s\} = \{a_1, \ldots, a_{m-1}\}$. Then $(a_1, C(a_1); \ldots; a_{m-1}, C(a_{m-1}); s)$ is an f.e.p.

(ii) $A^* \neq \emptyset$ and $s \notin A^*$.
Let $y \in A^*$ and let $C_y \subseteq \{t \in \Omega \mid z\,\mathbf{R}(t)\,y$ for all $z \in A\}$ satisfy $\lambda(C_y) = \beta(y)$. By the induction hypothesis for $\mathbf{R}^{\Omega \setminus C_y}{}_{|A \setminus \{y\}}$ there exists an f.e.p. $(x_1, C_1; \ldots; x_{m-1})$ for the restricted profile. Then $(y, C_y; x_1, C_1; \ldots; x_{m-1})$ is an f.e.p. for $\mathbf{R}$.

(iii) $s \in A^*$.
This case follows from Lemma 11.6.2. □

We shall use the existence of feasible elimination procedures established in Theorem 11.6.5 to derive the existence of an interesting class of ESC social choice functions, similarly as we did in Chapter 9. Let $\mathbf{R} \in \rho$. Call $x \in A$ **R**-*maximal* if there exists an f.e.p. $(x_1, C_1; \ldots; x_m)$ with respect to $\mathbf{R}$ such that $x = x_m$. Further, denote

$$M(\mathbf{R}) = \{x \in A \mid x \text{ is } \mathbf{R}\text{-maximal}\}.$$

Thus, $M(\mathbf{R}) \neq \emptyset$ for all $\mathbf{R} \in \rho$ if Assumption 11.6.3 holds. The following observation concerning $M(\cdot)$ will be very useful below.

*Remark 11.6.6.* Let $\mathbf{R} \in \rho$ and let $x \in A \setminus \{s\}$ satisfy

$$\lambda(\{t \in \Omega \mid y\,\mathbf{R}(t)\,x \text{ for all } y \in A\}) \geq \beta(x).$$

Then $x \notin M(\mathbf{R})$. This is so since $\lambda(\bigcup_{y \in A \setminus \{x\}} \{t \in \Omega \mid A \setminus \{y\} \mathbf{R}(t) y\}) \leq \lambda(\Omega) - \beta(x)$ and $s$ has to be eliminated strictly in an f.e.p.

**Theorem 11.6.7.** *Let Assumption 11.6.3 hold. Let the social choice function $F : \rho \to A$ be a selection from $M(\cdot)$, that is, $F(\mathbf{R}) \in M(\mathbf{R})$ for every $\mathbf{R} \in \rho$. Then $F$ is exactly and strongly consistent.*

*Proof.* Let $\mathbf{R} \in \rho$ and $x = F(\mathbf{R})$. Then there exists an f.e.p. $(x_1, C_1; \ldots; x_{m-1}, C_{m-1}; x)$ with respect to $\mathbf{R}$. Choose $\mathbf{Q} \in \rho$ that satisfies $y\,\mathbf{Q}(t)\,x_j$ for all $t \in C_j$, $y \in A$, and $j = 1, \ldots, m-1$. We claim that $F(\mathbf{Q}) = F(\mathbf{R})$ and that $\mathbf{Q}$ is a strong equilibrium of the game $(F, \mathbf{R})$. We distinguish the following cases.

(i) $x = s$.
By Remark 11.6.6, $F(\mathbf{Q}) = s$. Now assume, on the contrary, that $\mathbf{Q}$ is not a strong equilibrium of $(F, \mathbf{R})$. Then there exist $S \in \Sigma_+$ and $\mathbf{V}^S \in \rho^S$ such that $F(\mathbf{Q}^{\Omega \setminus S}, \mathbf{V}^S) = y$, $y \neq s$, and $y\,\mathbf{R}(t)\,s$ for all $t \in S$. Let $y = x_j$ for some $1 \leq j \leq m-1$. Then $S \cap C_j = \emptyset$ because $s\,\mathbf{R}(t)\,x_j$ for all $t \in C_j$. Hence, by Remark 11.6.6, $F(\mathbf{Q}^{\Omega \setminus S}, \mathbf{V}^S) \neq x_j$, which is the desired contradiction.

(ii) $x \neq s$.

Then $s = x_{j_0}$ for some $j_0 \leq m-1$. Thus, by definition of an f.e.p., $\lambda(C_{j_0}) > \beta(s)$. Hence, it is not possible to eliminate all $x' \neq s$ in an f.e.p. with respect to $\mathbf{Q}$, and therefore $F(\mathbf{Q}) \neq s$. By Remark 11.6.6 applied to all $x' \in A \setminus \{x, s\}$, $F(\mathbf{Q}) = x$. The proof that $\mathbf{Q}$ is a strong equilibrium of $(F, \mathbf{R})$ is analogous to that in case (i), observing that a profitable deviation from $\mathbf{Q}$ can never result in $s$ since $\lambda(C_{j_0}) > \beta(s)$ and, therefore, it is not possible to eliminate all alternatives in $A \setminus \{s\}$ in an f.e.p. with respect to $\mathbf{Q}$. □

We conclude this section with some observations which relate Theorem 11.6.7 to the preceding sections.

Let $\hat{F}$ be an anonymous selection from $M(\cdot)$. For instance, for every $\mathbf{R} \in \rho$ select the maximal element in $M(\mathbf{R})$ according to a fixed order $R_0 \in L$. By Theorem 11.6.7, $\hat{F}$ is an anonymous ESC social choice function, and therefore its associated effectivity function $E^{\hat{F}}$ is characterized by blocking coefficients (say) $\hat{\beta}(B)$ for $B \in P_0(A)$. Since alternatives assigned by $\hat{F}$ result from feasible elimination procedures with weights $\beta(a)$ $(a \in A)$, it is easy to check that $\hat{\beta}(a) = \beta(a)$ for every $a \in A$, and that $\{s\}$ is an i-set whereas all other singleton sets are e-sets. By the results established in Section 11.5, it follows that a set $B \subseteq A$ is an i-set if and only if it contains $s$. Note that the effectivity function $E^{\hat{F}}$ is independent of the particular anonymous selection $\hat{F}$ chosen since it is completely determined by the weights $\beta(a)$ $(a \in A)$, and thus we can denote it by $\hat{E}$. Since, for all $\mathbf{R} \in \rho$ and for every element $x \in M(\mathbf{R})$ we can always find an anonymous selection choosing that particular element, Theorem 11.4.5 implies that $M(\mathbf{R}) \subseteq C(\hat{E}, \mathbf{R})$ for all $\mathbf{R} \in \rho$. We can also state this as $M(\mathbf{R}) \subseteq C(E, \mathbf{R})$ for all $\mathbf{R} \in \rho$, where $E$ is the effectivity function associated with the system $(\beta, \mathbf{e}, \mathbf{i})$ as above (cf. Section 11.5). In the next section we shall establish the converse inclusion $C(E, \mathbf{R}) \subseteq M(\mathbf{R})$.

## 11.7 Core and feasible elimination procedures

In this section we prove that for any anonymous ESC social choice function that has exactly one i-alternative, every element in the core of the associated effectivity function can be obtained by a feasible elimination procedure.

Let $(\beta; \mathbf{e}, \mathbf{i})$ be a system satisfying (11.10)–(11.12) with $\mathbf{i}$ containing exactly one singleton $\{s\}$ for some designated $s \in A$. Hence, $\beta(y) > 0$ for all $y \in A \setminus \{s\}$. Let $E$ be the associated effectivity function. Note that $\beta(\cdot)$ satisfies Assumption 11.6.3 and therefore $M(\mathbf{R}) \neq \emptyset$ for all $\mathbf{R} \in \rho$ by Theorem 11.6.5. As explained in the last paragraph of Section 11.6, we have $M(\mathbf{R}) \subseteq C(E, \mathbf{R})$ and in particular $C(E, \mathbf{R}) \neq \emptyset$ for every $\mathbf{R} \in \rho$.

Let $\mathbf{R} \in \rho$ and $x \in C(E, \mathbf{R})$. For every $y \in A \setminus \{x\}$ denote

$$S(y) = \{t \in \Omega \mid x\, \mathbf{R}(t)\, y\}.$$

As before for $B \in P_0(A)$ we denote $\beta(B) = \sum_{y \in B} \beta(y)$. Using (11.13), (11.14), and the definition of the core it is straightforward to derive, for all $z \in A$,

$$z \in C(E,\mathbf{R}) \Leftrightarrow \begin{cases} \lambda(\cup_{y \in B} S_z(y)) \geq \beta(B) \text{ for all } B \in P_0(A) \text{ with } s \notin B, z \notin B, \\ \lambda(\cup_{y \in B} S_z(y)) > \beta(B) \text{ for all } B \in P_0(A) \text{ with } s \in B, z \notin B, \end{cases} \tag{11.20}$$

where $S_z(y) = \{t \in \Omega \mid z\,\mathbf{R}(t)\,y\}$, so $S(y) = S_x(y)$. In particular, we have $\lambda(S(y)) \geq \beta(y)$ for all $y \neq x$, with strict inequality if $y = s$. Of course, the sets $S(y)$ need not be disjoint, but Theorem 11.7.1 below says that they can be shrunk in such a way that they become pairwise disjoint while maintaining the inequalities. This theorem is a continuous version of the discrete 'marriage theorem' (cf. Halmos and Vaughan, 1950), suitable for our context, in particular for deriving Theorem 11.7.2 below. The latter theorem says that core elements are maximal, and its proof follows the construction in the proof of Theorem 11.7.1.[5]

**Theorem 11.7.1.** *There exist pairwise disjoint measurable sets $C(y)$, $y \in A \setminus \{x\}$, such that (i) $C(y) \subseteq S(y)$ for every $y \in A \setminus \{x\}$; (ii) $\lambda(C(y)) \geq \beta(y)$ for all $y \neq x$ and $\lambda(C(s)) > \beta(s)$.*

*Proof.* We start by noting that, if $x \neq s$, we may increase $\beta(s)$ with a small $\varepsilon > 0$ and decrease $\beta(x)$ with the same amount (note that $\beta(x) > 0$). In this way, all inequalities in (11.20) still hold as weak inequalities and it is sufficient to prove (*ii*) in the theorem with only weak inequalities. Moreover, we may regard $x$ as the i-alternative instead of $s$. For the rest of the proof we assume that this is the case.

We prove the theorem by induction on $|A| = m \geq 2$. The case $m = 2$ is obvious, so we concentrate on the induction step for $m \geq 3$. We first make the following observation.

*Remark.* Suppose there exists a set $B^* \subseteq A \setminus \{x\}$ with $\emptyset \neq B^* \neq A \setminus \{x\}$, such that $\lambda(\cup_{y \in B^*} S(y)) = \beta(B^*)$. In this case, we can decompose our problem into two smaller problems to which we can apply the induction hypothesis, as follows.

(i) The problem with set of alternatives $B^* \cup \{x\}$, set of voters $\cup_{y \in B^*} S(y)$, blocking coefficients $\hat{\beta}(x) = 0$ and $\hat{\beta}(y) = \beta(y)$ unchanged for $y \in B^*$, and preferences $\mathbf{R}(t)_{|B^* \cup \{x\}}$ for $t \in \cup_{y \in B^*} S(y)$. Note that all inequalities as in (11.20) restricted to voters in $\cup_{y \in B^*} S(y)$ and alternatives in $B^* \cup \{x\}$ still hold, and that $\lambda(\cup_{y \in B^*} S(y)) = \beta(B^*) = \hat{\beta}(B^* \cup \{x\})$.

(ii) The problem with set of alternatives $A \setminus B^*$, set of voters $\Omega \setminus \cup_{y \in B^*} S(y)$, blocking coefficients unchanged, and preferences $\mathbf{R}(t)$ restricted to $A \setminus B^*$.

---

[5] For a proof of a slightly less general version of the continuous 'marriage theorem' see Hart and Kohlberg (1974, p. 171).

Note that all inequalities still hold since for any set $B \subseteq A \setminus (\{x\} \cup B^*)$ we have

$$\begin{aligned}\lambda(\cup_{y \in B} S(y) \setminus \cup_{\hat{y} \in B^*} S(\hat{y})) &= \lambda(\cup_{y \in B \cup B^*} S(y)) - \lambda(\cup_{y \in B^*} S(y)) \\ &= \lambda(\cup_{y \in B \cup B^*} S(y)) - \beta(B^*) \\ &\geq \beta(B \cup B^*) - \beta(B^*) \\ &= \beta(B).\end{aligned}$$

Furthermore, $\lambda(\Omega \setminus \cup_{y \in B^*} S(y)) = \beta(A \setminus B^*)$.

The required sets $C(y)$, $y \in A \setminus \{x\}$ are now obtained by applying the induction hypothesis to each subproblem.

We now proceed to the induction step. Let $m \geq 3$. We are done if there is a decomposition possible as in the Remark, so suppose there is none. Let $b \in A \setminus \{x\}$ and consider the set $S = S(b) \setminus \cup_{y \in A \setminus \{x,b\}} S(y)$, i.e., $S = \{t \in \Omega \mid y\, \mathbf{R}(t)\, x\, \mathbf{R}(t)\, b$ for all $y \neq x, b\}$. We distinguish two cases.

*Case 1*: $\lambda(S) \geq \beta(b)$. Since $x \in C(E, \mathbf{R})$, $0 \leq \lambda(S) \leq \beta(x) + \beta(b)$. Now take $C(b)$ equal to $S$, and apply the induction hypothesis to the problem with set of alternatives $A \setminus \{b\}$, set of voters $\Omega \setminus S$, blocking weights $\beta'$ unchanged except $\beta'(x) = \beta(x) - (\lambda(S) - \beta(b))$, and preferences equal to the original preferences restricted to $A \setminus \{b\}$.

*Case 2:* $\lambda(S) < \beta(b)$. We also know $\lambda(S(b)) > \beta(b)$ otherwise $\lambda(S(b)) = \beta(b)$ by (11.20), and we would have a decomposition as in the Remark with $B^* = \{b\}$. Now choose a measurable set $S^*$ satisfying $S \subseteq S^* \subseteq S(b)$ and $\lambda(S^*) = \beta(b)$ (this is possible by Lyapunov's Theorem). Consider the set of vectors

$$\{(\lambda\left(S^* \cup T \cup (\cup_{y \in B} S(y))\right))_{B \subsetneq A \setminus \{b,x\}} \mid \emptyset \subseteq T \subseteq S(b) \setminus S^*\}. \tag{11.21}$$

For $T = S(b) \setminus S^*$ and $B = \emptyset$ we have

$$\lambda(S^* \cup T \cup (\cup_{y \in B} S(y))) = \lambda(S(b)) > \beta(b) \tag{11.22}$$

and for $T = S(b) \setminus S^*$ and $B \subseteq A \setminus \{x, b\}$ arbitrary we have

$$\lambda(S^* \cup T \cup (\cup_{y \in B} S(y))) = \lambda(\cup_{y \in B \cup \{b\}} S(y)) \geq \beta(b) + \beta(B) \tag{11.23}$$

by (11.20). For $T = \emptyset$ and $B = \emptyset$ we have

$$\lambda(S^* \cup T \cup (\cup_{y \in B} S(y))) = \lambda(S^*) = \beta(b). \tag{11.24}$$

Now for $B \subseteq A \setminus \{x, b\}$ with $B \neq A \setminus \{x, b\}$ and $T \subseteq S(b) \setminus S^*$ consider the expression

$$\begin{aligned}\lambda(S^* \cup T \cup (\cup_{y \in B} S(y))) = \lambda(T) &+ \lambda(S^* \cup (\cup_{y \in B} S(y))) \\ &- \lambda(T \cap (S^* \cup (\cup_{y \in B} S(y)))).\end{aligned}$$

This is an affine function, with variable $T$, of two measures $\lambda(T)$ and $\lambda(T \cap (S^* \cup (\cup_{y \in B} S(y))))$. As $B$ varies on $\{B' \mid B' \subseteq A \setminus \{b, x\},\ B' \neq A \setminus \{b, x\}\}$ we obtain an affine combination of two vector measures. Hence, its range

(11.21) is compact and convex by Lyapunov's Theorem. By (11.22), (11.23), and (11.24), we can choose $T = T_0$ such that all inequalities in (11.23) are still valid but with at least one equality, say for $B_0$. Now set $S_0 = S^* \cup T_0$, and set $B^* = B_0 \cup \{b\}$. On $S(b) \setminus S_0$ change the preferences by shifting $b$ up so that it becomes preferred to $x$. Use the notation $\bar{S}(\cdot)$ for the $S(\cdot)$-sets in the new profile. Then all sets $S(y)$, $y \neq b$, remain unchanged, i.e., $\bar{S}(y) = S(y)$, whereas $S(b)$ changes to $\bar{S}(b) = S_0$. Then, for this *new* profile, we have $\beta(B^*) = \beta(b) + \beta(B_0) = \lambda(S_0 \cup (\cup_{y \in B_0} S(y))) = \lambda(\cup_{y \in B^*} \bar{S}(y))$. The problem with the new profile is decomposable according to the Remark. Applying the Remark, we obtain the desired sets: in particular, the resulting set $C(b)$ is a subset of $\bar{S}(b) = S_0$ and therefore of $S(b)$. This concludes the proof of the theorem. □

Still under the assumptions made at the beginning of this section we proceed to show that $x$ is a maximal alternative, i.e., $x \in M(\mathbf{R})$. We first attach a precise and formal meaning to the expression 'bottom alternative': we call $b \in A$ a *bottom alternative* of $\mathbf{R}$ if the set $\hat{S}(b) = \{t \in \Omega \mid y\, \mathbf{R}(t)\, b \text{ for all } y \in A\}$ has measure $\lambda(\hat{S}(b)) \geq \beta(b)$, with strict inequality sign for $b = s$. Observe that there is always a bottom alternative since $\sum_{a \in A} \beta(a) = \lambda(\Omega)$. Obviously, $x$ is not a bottom alternative since it is in the core $C(E, \mathbf{R})$.

We have the following result.[6]

**Theorem 11.7.2.** *Alternative $x$ is $\mathbf{R}$-maximal, that is, $x \in M(\mathbf{R})$. In particular, if $b$ is a bottom alternative of $\mathbf{R}$, then there is an f.e.p. $(b, C_b; y_1, C_1; \ldots; y_{m-2}, C_{m-2}; x)$.*

*Proof.* Let $b$ be a bottom alternative. If $b = s$ we slightly increase the blocking coefficient of $b$ (as in the beginning of the proof of Theorem 11.7.1) so that we still have $\lambda(\hat{S}(b)) \geq \beta(b)$. (This has the advantage that in what follows it is sufficient to consider blocking with weak inequalities.)

The proof is by induction on $m = |A|$. For $m = 2$ the result is again obvious. Let $m \geq 3$.

(i) First suppose that the problem is decomposable into two subproblems with sets of alternatives $\{x\} \cup B^*$ and $A \setminus B^*$ as in the proof of Theorem 11.7.1, and with $b \in B^*$. Note that all voters in the problem with $A \setminus B^*$ rank $B^*$ above $x$. By the induction hypothesis, each of the subproblems has an f.e.p. leading to $x$, with the one in the first subproblem starting with $b$. Let $|B^*| = k$, let $(b, C_b; y_1, C_1; \ldots; y_{k-1}, C_{k-1}; x)$ be an f.e.p. in the problem with $\{x\} \cup B^*$ and let $(x_1, \hat{C}_1; \ldots; x_{m-k-1}, \hat{C}_{m-k-1}; x)$ be an f.e.p. in the problem with $A \setminus B^*$. Then

$$(b, C_b; x_1, \hat{C}_1; \ldots; x_{m-k-1}, \hat{C}_{m-k-1}; y_1, C_1; \ldots; y_{k-1}, C_{k-1}; x)$$

is an f.e.p. for the original problem.

---

[6] The analogous result to Theorem 11.7.2 for the case with finitely many voters is Theorem 9.3.6. The proof of the latter theorem has benefitted from the analysis in this chapter.

(ii) Next, suppose the problem is not decomposable in this way. As in the proof of Theorem 11.7.1 let $S = S(b) \setminus \bigcup_{y \in A \setminus \{x,b\}} S(y)$ and distinguish two cases as there. In Case 1, $\lambda(S) \geq \beta(b)$, we take again $C(b) = S$, observing that $S \subseteq \hat{S}(b)$. Applying the induction hypothesis, we let $(y_1, C_1; \ldots; y_{m-2}, C_{m-2}; x)$ be an f.e.p. in the problem with set of alternatives $A \setminus \{b\}$, then $(b, C_b; y_1, C_1; \ldots; y_{m-2}, C_{m-2}; x)$ is as desired.

In Case 2, we proceed again as in the proof of Theorem 11.7.1 but we make sure that $S_0$ there is chosen in such a way that $\lambda(S_0 \cap \hat{S}(b)) \geq \beta(b)$. This is possible since $S \subseteq \hat{S}(b) \subseteq S(b)$ and so we can choose $S^*$ (which is a subset of $S_0$ by construction) such that $S^* \subseteq \hat{S}(b)$. We have now again a decomposition as in (i) of this proof: since $b$ is eliminated first, shifting $b$ over $x$ in the original preferences of voters in $S(b) \setminus S_0$ does not change the restriction of these preferences to $A \setminus B^*$. □

We conclude this section by summarizing the main results of Sections 11.6 and 11.7 in the following corollary.

**Corollary 11.7.3.**

(*i*) *Let $F$ be an anonymous ESC social choice function. Suppose that the associated effectivity function $E$ has exactly one i-alternative. Then $C(E, \cdot) = M(\cdot)$ and $F$ is a selection from this set.*

(*ii*) *Let $(\beta; \mathbf{e}, \mathbf{i})$ be a system satisfying (11.10)–(11.12) such that $\mathbf{i}$ contains exactly one singleton. Then, for the associated effectivity function $E$, $C(E, \cdot) = M(\cdot)$, and any anonymous selection from this set is an anonymous ESC social choice function.*

## 11.8 Positive blocking coefficients

A natural question is whether Corollary 11.7.3 can be extended to general systems $(\beta; \mathbf{e}, \mathbf{i})$. We have already remarked that if there are two or more i-alternatives, then a feasible elimination procedure may fail to exist. On the other hand, Example 11.4.2 (or 11.5.3) shows that an anonymous ESC social choice function may generate more than one i-alternative. In this section we show that if an anonymous ESC social choice function generates only positive blocking coefficients, then there can be at most one i-alternative. In other words, Corollary 11.7.3 provides a complete characterization of anonymous ESC social choice functions if we require all blocking coefficients to be positive.

Let $E$ be the effectivity function associated with a system $(\beta; \mathbf{e}, \mathbf{i})$, satisfying (11.10)–(11.12).

**Definition 11.8.1.** $E$ satisfies $D(k)$, where $1 \leq k \leq m-2$, if there exist *no* partitions $\{x_1\}, \ldots, \{x_k\}, C_1, C_2$ of $A$ and $S_1, \ldots, S_k, T_1, T_2$ of $\Omega$, $S_1, \ldots, S_k, T_1, T_2 \in \Sigma_+$, such that[7]

(i) $\lambda(S_i) = \beta(x_i)$ for $i = 1, \ldots, k$, and $x_1, \ldots, x_k$ are e-alternatives;
(ii) $\lambda(T_i) = \beta(C_i)$ for $i = 1, 2$, and $C_1$ and $C_2$ are i-sets.

The following theorem is a counterpart of similar results for the case of finitely many voters, see Section 10.4 in particular. Its proof is deferred until the end of this section.

**Theorem 11.8.2.** *Let $F : \rho \to A$ be an anonymous ESC social choice function, and let $(\beta; \mathbf{e}, \mathbf{i})$ be the associated system. Suppose that $\beta(a) > 0$ for all $a \in \mathbf{i}$. Then $E = E^F$ satisfies $D(k)$ for all $1 \leq k \leq m-2$.*

We now have:

**Corollary 11.8.3.** *Let $F : \rho \to A$ be an anonymous ESC social choice function that generates only positive blocking coefficients. Then there is exactly one i-alternative.*

*Proof.* Clearly, by (11.11) and (11.12), there must be at least one i-alternative: if all alternatives were e-alternatives then repeated application of (11.12) would give a violation of (11.11). Also, there must be at least one e-alternative: if not, then $A \setminus \{x\}$ would be an e-set for each $x \in A$ by (11.11), hence $A \setminus \{x, y\} = A \setminus \{x\} \cap A \setminus \{y\}$ would be an e-set for all $x, y \in A$ by (11.12), and so on and so forth, implying that all singletons would be e-sets, a contradiction.

Suppose that there are two different i-alternatives $x, y$ in the associated system. Let $\{x_1\}, \ldots, \{x_k\}$ be the e-singletons, hence $1 \leq k \leq m-2$. Define $C_1 = \{x\}$ and $C_2 = \{z \in A \mid z \neq x, \{z\} \in \mathbf{i}\}$. Then $C_2$ is an i-set, which can be seen as follows. Write $C_2 = \{y_1, \ldots, y_\ell\}$, where $\ell \geq 1$. If $C_2$ were an e-set, then also $C_2 \setminus \{y_\ell\} = C_2 \cap A \setminus \{y_\ell\}$ would be an e-set by (11.12). Hence, $C_2 \setminus \{y_\ell, y_{\ell-1}\} = C_2 \setminus \{y_\ell\} \cap A \setminus \{y_{\ell-1}\}$ is an e-set, and so on and so forth, until we obtain that $\{y_1\}$ is an e-set, which is a contradiction.

Now choose a partition of $\Omega$ as in Definition 11.8.1, so that $D(k)$ is violated. This contradicts Theorem 11.8.2. □

The combination of Corollaries 11.7.3 and 11.8.3 yields an almost complete characterization of anonymous ESC social choice functions. The case where there is more than one i-alternative – so that at least one i-alternative has zero blocking coefficient – is still open.

### *Proof of Theorem 11.8.2*

We start with the following observation.

---

[7] Recall from Section 11.3 that elements of a partition have positive measure by definition.

**Lemma 11.8.4.** *Let $F : \rho \to A$ be an ESC SCF. Then there exist no partitions $S_1, \ldots, S_p$ of $\Omega$ and $B_1, \ldots, B_p$ of $A$ (where $p \geq 2$) such that $A \setminus B_i \notin E^F(S_i)$ for all $i = 1, \ldots, p$.*

*Proof.* Assume, on the contrary, that there exist partitions as in the lemma. Consider the following profile:

| $S_1$ | $S_2$ | $\cdots$ | $S_p$ |
|---|---|---|---|
| $B_2$ | $B_3$ | $\cdots$ | $B_1$ |
| $B_3$ | $B_4$ | $\cdots$ | $B_2$ |
| $\vdots$ | $\vdots$ | | $\vdots$ |
| $B_p$ | $B_1$ | $\cdots$ | $B_{p-1}$ |
| $B_1$ | $B_2$ | $\cdots$ | $B_p$ |

By maximality (Theorem 11.4.5) of $E^F$, we have $B_i \in E^F(\Omega \setminus S_i)$ for every $i = 1, \ldots, p$. Hence, the alternatives in $B_2$ are blocked by $\Omega \setminus S_1$, the alternatives in $B_3$ by $\Omega \setminus S_2$, etc., so that $C(E^F, \mathbf{R}) = \emptyset$. But this contradicts stability (Theorem 11.4.5) of $E^F$. □

**Proof of Theorem 11.8.2** The proof is by induction on $k$.

(1) $k = 1$. Assume, on the contrary, that there are partitions $\{x_1\}$, $C_1$, $C_2$ of $A$ and $S_1$, $T_1$, $T_2$ of $\Omega$, satisfying (i) and (ii) in Definition 11.8.1. Let $S_1 = S^1 \cup S^2$ with $S^1 \cap S^2 = \emptyset$ and $\lambda(S^1) = \lambda(S^2)$. Consider the following profile $\mathbf{R}$:

| $S^1$ | $S^2$ | $T_1$ | $T_2$ |
|---|---|---|---|
| $C_1$ | $C_2$ | $x_1$ | $x_1$ |
| $C_2$ | $C_1$ | $C_2$ | $C_1$ |
| $x_1$ | $x_1$ | $C_1$ | $C_2$ |

Since $S_1$ can block $x_1$ (i.e., $A \setminus \{x_1\} \in E(S_1)$), stability of $E^F$ (Theorem 11.4.5) implies $F(\mathbf{R}) \neq x_1$. Without loss of generality $F(\mathbf{R}) \in C_1$. Let $\mathbf{Q}$ be a strong Nash equilibrium of $(F, \mathbf{R})$ with $F(\mathbf{R}) = F(\mathbf{Q})$. We distinguish the following possibilities.

(1.1) There exists $y \in C_2$ such that $\lambda(\{t \in S_1 \mid x_1 \, \mathbf{Q}(t) \, y\}) > 0$. Choose $S^3 \subseteq \{t \in S_1 \mid x_1 \, \mathbf{Q}(t) \, y\}$ such that $0 < \lambda(S^3) < \min_{a \in A} \beta(a)$. Define the $T_1 \cup T_2$-profile $\mathbf{P}$ by

$$x_1 \, \mathbf{P}(t) \, y \, \mathbf{P}(t) \, A \setminus \{x_1, y\} \quad \text{for all } t \in T_1 \cup T_2.$$

By considering the partitions $S_1 \setminus S^3$, $S^3$, $T_1 \cup T_2$, and $\{x_1\}$, $\{y\}$, $A \setminus \{x_1, y\}$, it follows from Lemma 11.8.4 that $T_1 \cup T_2$ blocks $A \setminus \{x_1, y\}$. Hence, $F(\mathbf{Q}^{S_1}, \mathbf{P}^{T_1 \cup T_2}) \in \{x_1, y\}$. As $T_1 \cup T_2 \cup S^3$ is effective for $x_1$, and $F(\mathbf{Q}^{S_1}, \mathbf{P}^{T_1 \cup T_2}) \in C(E, (\mathbf{Q}^{S_1}, \mathbf{P}^{T_1 \cup T_2}))$, we have $F(\mathbf{Q}^{S_1}, \mathbf{P}^{T_1 \cup T_2}) = x_1$. Thus, $T_1 \cup T_2$ has improved upon $F(\mathbf{Q})$, which is a contradiction.

(1.2) $C_2 \, \mathbf{Q}(t) \, x_1$ for all $t \in S_1$.
Consider the $T_1 \cup S^2$-profile $\mathbf{P}$ defined by

$C_2\,\mathbf{P}(t)\,x_1\,\mathbf{P}(t)\,C_1$ for all $t \in T_1 \cup S^2$.

Since $\lambda(T_1 \cup S^2) > \lambda(T_1) = \beta(C_1)$, $T_1 \cup S^2$ blocks $C_1$. Therefore, $F(\mathbf{Q}^{T_2 \cup S^1}, \mathbf{P}^{T_1 \cup S^2}) \notin C_1$. Suppose $F(\mathbf{Q}^{T_2 \cup S^1}, \mathbf{P}^{T_1 \cup S^2}) = x_1$. Note that, in the profile $(\mathbf{Q}^{T_2 \cup S^1}, \mathbf{P}^{T_1 \cup S^2})$, both $T_1$ and $S_1$ prefer $C_2$ over $x_1$. Moreover, $\lambda(T_1 \cup S_1) = \beta(C_1) + \beta(x_1)$ and $C_1 \cup \{x_1\}$ is an e-set, because $C_2$ is an i-set; therefore, $T_1 \cup S_1$ blocks $C_1 \cup \{x_1\}$. This contradicts $F(\mathbf{Q}^{T_2 \cup S^1}, \mathbf{P}^{T_1 \cup S^2}) = x_1$ and, hence, $F(\mathbf{Q}^{T_2 \cup S^1}, \mathbf{P}^{T_1 \cup S^2}) \in C_2$. But this contradicts the fact that $\mathbf{Q}$ is a strong Nash equilibrium in $(F, \mathbf{R})$.

(2) Let $1 < k \leq m - 2$ and assume $D(1), \ldots, D(k-1)$. We shall prove $D(k)$.

Assume, on the contrary, that there exist partitions $\{x_1\}, \ldots, \{x_k\}, C_1, C_2$ of $A$ and $S_1, \ldots, S_k, T_1, T_2$ of $\Omega$, satisfying (i) and (ii) in Definition 11.8.1. If $C_1 \cup \{x_k\}$ is an i-set, then we obtain a contradiction to $D(k-1)$. Otherwise, $A \backslash (C_1 \cup \{x_k\})$ is an i-set. Then consider the partitions $\{x_k\}, C_1, A \backslash (C_1 \cup \{x_k\})$ of $A$, and $S_k, T_1, \Omega \setminus (S_k \cup T_1)$ of $\Omega$: this implies a contradiction to $D(1)$. □

## 11.9 Notes and comments

Most of the results of this chapter first appeared in Peleg and Peters (2006). The extension of the Gibbard-Satterthwaite theorem to the continuum voter case first appeared in the working paper version of Peleg and Peters (2006). There, it is also shown that in this model an effectivity function is maximal and stable if and only if it can be represented by a strongly consistent game form. See Propositions 5.2.4 and 5.2.6 and Theorem 5.3.2, or Moulin and Peleg (1982), for the case with finitely many voters.

# References


1. Abdou, J. (1988): "Neutral Veto Correspondence with a Continuum of Alternatives," *International Journal of Game Theory*, 17, 135–164.
2. Abdou, J. (1991): "Maxmin and Minmax for Coalitional Game Forms," *Games and Economic Behavior*, 3, 267–277.
3. Abdou, J. (1995): "Nash and Strongly Consistent Two-Player Game Forms," *International Journal of Game Theory*, 24, 345–356.
4. Abdou, J., and H. Keiding (1991): *Effectivity Functions in Social Choice*. Kluwer Academic Publishers, Dordrecht.
5. Abdou, J., and H. Keiding (2003): "On Necesary and Sufficient Conditions for Solvability of Game Forms," *Mathematical Social Sciences*, 46, 243–260.
6. Abreu, D., and A. Sen (1991): "Virtual Implementation in Nash Equilibrium," *Econometrica*, 59, 997–1021.
7. Arrow, K.J. (1951, 1963) *Social Choice and Individual Values*. Wiley, New York.
8. Arrow, K.J. (1967): "Values and Collective Decision-Making," in: P. Laslett and W.G. Runciman (eds.), *Philosophy, Politics and Society*. Basil Blackwell Oxford UK.
9. Aumann, R.J. (1967): "A Survey of Cooperative Games without Side Payments," in: *Essays in Mathematical Economics*, ed. by M. Shubik, 3–27. Princeton University press, Princeton NJ.
10. Barberà, S. (2001): "An Introduction to Strategy-Proof Social Choice Functions," *Social Choice and Welfare*, 18, 619–653.
11. Barberà, S., B. Dutta, and A. Sen (2001): "Strategy-Proof Social Choice Correspondences," *Journal of Economic Theory*, 101, 374–394.
12. Batteau, P., J.-M. Blin, and B. Monjardet (1981): "Stability of Aggregation Procedures, Ultrafilters, and Simple Games," *Econometrica*, 49, 527–534.
13. Black, D. (1948): "On the Rationale of Group-Decision-Making," *Journal of Political Economy*, 56, 23–34.
14. Blin, J.-M., and M.A. Satterthwaite (1976): "Strategy-Proofness and Single-Peakedness," *Public Choice*, 26, 51–58.
15. Borm, P.E.M., and S.H. Tijs (1992): "Strategic Claim Games Corresponding to an NTU-Game," *Games and Economic Behavior*, 4, 58–71.
16. Brams, S.J., and P.C. Fishburn (1983): *Approval Voting*. Birkhauser, Boston MA.
17. Buchanan, J.M., and G. Tullock (1962): *The Calculus of Consent, Logical Foundations of Constitutional Democracy*. University of Michigan Press, Ann Arbor.
18. Dummett, M., and R. Farquharson (1961): "Stability in Voting," *Econometrica*, 29, 33–44.
19. Dunford, N., and J.T. Schwartz (1988): *Linear Operators, Part I: General Theory*. Wiley Classic Library, New York.

20. Dutta, B. (1984): "Effectivity Functions and Acceptable Game Forms," *Econometrica*, 52, 1151–1166.
21. Dutta, B., and P.K. Pattanaik (1978): "On Nicely Consistent Voting Systems," *Econometrica*, 78, 163–170.
22. Feldman, A.M. (1980): "Strongly Manipulable Multi-Valued Collective Choice Rules," *Public Choice*, 35, 503–509.
23. Feldman, A.M., and R. Serrano (2005): *Welfare Economics and Social Choice Theory.* 2nd edition, Springer-Verlag, Berlin Heidelberg New York.
24. Fishburn, P.C. (1972): "Even-Chance Lotteries in Social Choice Theory," *Theory and Decision*, 3, 18–40.
25. Friedman, M. (1962): *Capitalism and Frieedom.* Chicago University Press, Chicago (2002 edition).
26. Gaertner, W., P.K. Pattanaik, and K. Suzumura (1992): "Individual Rights Revisited," *Economica*, 59, 161–177.
27. Gärdenfors, P. (1981): "Rights, Games, and Social Choice," *Noûs*, 15, 341–356.
28. Gibbard, A. (1973): "Manipulation of Voting Schemes: A General Result," *Econometrica*, 41, 587–602.
29. Gibbard, A. (1974): "A Pareto-consistent Libertarian Claim," *Journal of Economic Theory*, 7, 388–410.
30. Gurvich, V.A. (1989): "Equilibrium in Pure Strategies," *Soviet Mathematics Doklady*, 38, 597–602.
31. Halmos, P.R., and H.E. Vaughan (1950): "The Marriage Problem," *American Journal of Mathematics*, 72, 214–215.
32. Harsanyi, J. (1973): "Games with Randomly Disturbed Payoffs: A New Rationale for Mixed Strategy Equilibrium Points," *International Journal of Game Theory*, 2, 1–23.
33. Hart, S., and E. Kohlberg (1974): "Equally Distributed Correspondences," *Journal of Mathematical Economics*, 1, 167–174.
34. Hausdorff, F. (1962): *Set Theory.* Chelsea Publishing Company, New York.
35. Hildenbrand, W. (1974): *Core and Equilibria of a Large Economy.* Princeton University Press, Princeton.
36. Holzman, R. (1986a): "The Capacity of a Committee," *Mathematical Social Sciences*, 12, 139–157.
37. Holzman, R. (1986b): "On Strong Representations of Games by Social Choice Functions," *Journal of Mathematical Economics*, 15, 39–57.
38. Hurwicz, L., and D. Schmeidler (1978): "Construction of Outcome Functions Guaranteeing Existence and Pareto Optimality of Nash Equilibria," *Econometrica*, 46, 1447–1474.
39. Ishikawa, S., and K. Nakamura (1980): "Representations of Characteristic Function Games by Social Choice Functions," *International Journal of Game Theory*, 9, 191–199.
40. Kanger, S., and H. Kanger (1972): "Rights and Parliamentarism," in: Olson, R.E., Paul, A.M. (eds) *Contemporary Philosophy in Scandinavia.* The John Hopkins Press, Baltimore, 213–236.
41. Keiding, H. (1985): "Necessary and Sufficient Conditions for Stability of Effectivity Functions," *International Journal of Game Theory*, 14, 93–101.
42. Keiding, H., and B. Peleg (2006a): "Binary Effectivity Rules," *Review of Economic Design*, 10, 167–181.
43. Keiding, H., and B. Peleg (2006b): "On the Continuity of Representations of Effectivity Functions," *Journal of Mathematical Economics*, 42, 827–842.
44. Kelley, J.L. (1955): *General Topology.* Springer-Verlag, New York Berlin Heidelberg.
45. Kelly, J.S. (1988): "Minimal Manipulability and Local Strategy-Proofness," *Social Choice and Welfare*, 5, 81–85.
46. Kirman, A., and D. Sondermann (1972): "Arrow's Theorem, Many Agents, and Invisible Dictators," *Journal of Economic Theory*, 3, 267–277.

47. KLEIN, E., AND A.C. THOMPSON (1984): *Theory of Correspondences.* Wiley, New York.
48. MAS-COLELL, A., AND H.F. SONNENSCHEIN (1972): "General Possibility Theorems for Group Decisions," *Review of Economic Studies*, 39, 185–192.
49. MASKIN, E. (1999): "Nash Equilibrium and Welfare Optimality," *The Review of Economic Studies*, 66, 23–38.
50. MAUS, S., H. PETERS, AND T. STORCKEN (2007): "Minimal Manipulability: Unanimity and Nondictatorship," *Journal of Mathematical Economics*, 43, 675–691.
51. MOULIN, H. (1980): "On Strategy-Proofness and Single Peakedness," *Public Choice*, 35, 437–455.
52. MOULIN, H., AND B. PELEG (1982): "Cores of Effectivity Functions and Implementation Theory," *Journal of Mathematical Economics*, 10, 115–145.
53. MULLER, E., AND M.A. SATTERTHWAITE (1977): "The Equivalence of Strong Positive Association and Strategy-Proofness, *Journal of Economic Theory*, 14, 412–418.
54. NASH, J.F. (1951): "Non-cooperative games," *Annals of Mathematics*, 54, 286–295.
55. OREN, I (1981): "The Structure of Exactly Strongly Consistent Social Choice Functions," *Journal of Mathematical Economics*, 8, 207–220.
56. PATTANAIK, P.K. (1976): "Counter-Threats and Strategic Manipulation under Voting Schemes," *The Review of Economic Studies*, 43, 11–18.
57. PELEG, B. (1978a): "Consistent Voting Systems," *Econometrica*, 46, 153–161.
58. PELEG, B. (1978b): "Representation of Simple Games by Social Choice Functions," *International Journal of Game Theory*, 7, 81–94.
59. PELEG, B. (1984): *Game Theoretic Analysis of Voting in Committees.* Cambridge (UK), Cambridge University Press.
60. PELEG, B. (1998): "Effectivity Functions, Game Forms, Games, and Rights," *Social Choice and Welfare*, 15, 67–80.
61. PELEG, B. (2004): "Representation of Effectivity Functions by Acceptable Game Forms: A Complete Characterization," *Mathematical Social Sciences*, 43, 275–287.
62. PELEG, B., AND H. PETERS (2006): "Consistent Voting Systems with a Continuum of Voters," *Social Choice and Welfare*, 27, 477–492.
63. PELEG, B., AND H. PETERS (2009): "Nash Consistent Representation of Effectivity Functions Through Lottery Models," *Games and Economic Behavior*, 65, 503–515.
64. PELEG, B., H. PETERS, AND T. STORCKEN (2002): "Nash Consistent Representation of Consitutions: A Reaction to the Gibbard Paradox," *Mathematical Social Sciences*, 43, 267–287.
65. PELEG, B., AND A.D. PROCACCIA (2007): "Mediators Enable Truthful Voting," DP 451 Center for the Study of Rationality, The Hebrew University of Jerusalem.
66. PETERS, H. (2008): *Game Theory: A Multi-Leveled Approach.* Springer-Verlag, Berlin Heidelberg.
67. POLISHCHUK, I. (1978): "Monotonicity and Uniqueness of Consistent Voting Systems," Center for Research in Mathematical Economics and Game Theory, Hebrew University of Jerusalem.
68. ROY, S., H. PETERS, AND T. STORCKEN (2009): "On the Manipulability of Approval Voting," working paper, Department of Quantitative Economics, Maastricht University.
69. SATTERTHWAITE, M. (1975): "Strategy-proofness and Arrow's Conditions: Existence and Correspondence Theorems for Voting Procedures and Social Welfare Functions," *Journal of Economic Theory*, 10, 187–217.
70. SCHELLING, T.C. (1960): *The Strategy of Conflict.* Harvard University Press, Cambridge MA.
71. SEN, A.K. (1970): "The Impossibility of a Paretian Liberal," *Journal of Political Economy*, 78, 152–157.
72. SEN, A.K. (1997): "Individual Preference as the Basis of Social Choice," in: K.J. Arrow, A.K. Sen, and K. Suzumura (eds.), *Social Choice Re-Examined.* Macmillan, London.
73. VON NEUMANN, J., AND O. MORGENSTERN (1944): *Theory of Games and Economic Behavior.* Princeton University Press, Princeton NJ.

# Author Index

# Subject Index

Italic page numbers indicate definitions of subjects.

FSC
www.fsc.org
MIX
Papier | Fördert
gute Waldnutzung
FSC® C083411

Kolibri 360 GmbH
Ferdinand-Jühlke-Straße 7
99095 Erfurt, Deutschland
produktsicherheit@kolibri360.de